SCIENCE NOTEBOOKS

in Student-Centered Classrooms

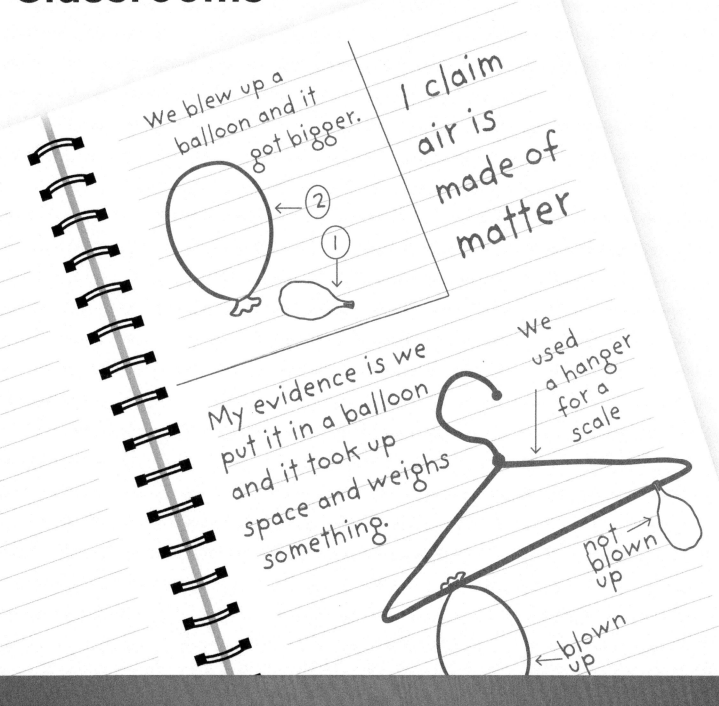

We blew up a balloon and it got bigger.

I claim air is made of matter

My evidence is we put it in a balloon and it took up space and weighs something.

We used a hanger for a scale

not → blown up

← blown up

SCIENCE NOTEBOOKS
in Student-Centered Classrooms

We blew up a balloon and it got bigger.

I claim air is made of matter

② ①

My evidence is we put it in a balloon and it took up space and weighs something.

We used a hanger for a scale

not blown up

blown

Jessica Fries-Gaither

nsta Press
National Science Teaching Association
Arlington, Virginia

ART AND DESIGN
Will Thomas Jr., Director

PRINTING AND PRODUCTION
Catherine Lorrain, Director

NATIONAL SCIENCE TEACHING ASSOCIATION
1840 Wilson Blvd., Arlington, VA 22201
www.nsta.org/store
For customer service inquiries, please call 800-277-5300.

NSTA is committed to publishing material that promotes the best in inquiry-based science education. However, conditions of actual use may vary, and the safety procedures and practices described in this book are intended to serve only as a guide. Additional precautionary measures may be required. NSTA and the authors do not warrant or represent that the procedures and practices in this book meet any safety code or standard of federal, state, or local regulations. NSTA and the authors disclaim any liability for personal injury or damage to property arising out of or relating to the use of this book, including any of the recommendations, instructions, or materials contained therein.

Library of Congress Cataloging-in-Publication Data
Names: Fries-Gaither, Jessica, 1977- author.
Title: Science notebooks in student-centered classrooms / by Jessica Fries-Gaither.
Description: Arlington, VA : National Science Teaching Association, [2022] |
    Includes bibliographical references and index.
Identifiers: LCCN 2021045020 (print) | LCCN 2021045021 (ebook) | ISBN 9781681407074 (paperback) |
    ISBN 9781681407081 (pdf)
Subjects: LCSH: Science--Study and teaching--United States. | Student-centered learning--Methodology.
Classification: LCC Q183.3.A1 F74 2022 (print) | LCC Q183.3.A1 (ebook) |
    DDC 507.1/073--dc23/eng/20211006
LC record available at *https://lccn.loc.gov/2021045020*
LC ebook record available at *https://lccn.loc.gov/2021045021*

CONTENTS

ACKNOWLEDGMENTS

I thank my parents for fostering my love of learning and, in particular, my passion for reading, writing, and science. The weekly library visits, walks in our local metro parks, time to splash and tromp around in nature, and summer workshops at our science museum all set me on this path.

I am grateful for my husband's constant support of my writing endeavors.

Pat Brown's thoughtful and insightful critiques of my early drafts and revisions helped me immensely throughout the writing process. His encouraging words gave me confidence when I questioned whether I had the expertise to write this book at all.

I appreciate Claire Reinburg for suggesting this as the topic of my next book, Kate Soriano for her enthusiastic help with revisions, and the NSTA Press team for their ongoing support.

Heartfelt thanks go to Nichole Bondi and Ben Simon for their willingness to photograph a veritable mountain of student work.

Above all, I thank my students, past and present. Their curiosity, creativity, and joy inspire me daily.

ABOUT THE AUTHOR

Jessica Fries-Gaither is an experienced science educator and an award-winning author of books for students and teachers. She is currently chair of the science department and the science specialist for the Lower School at the Columbus School for Girls in Columbus, Ohio. Jessica is a reviewer for NSTA's elementary journal, *Science and Children,* and has served on several NSTA advisory boards. She presents at local, regional, and national conferences and teaches elementary preservice teachers in Notre Dame's Alliance for Catholic Education program. Jessica and her husband live in Columbus with their four dogs. She enjoys reading, cooking, and spending time outside.

INTRODUCTION

Imagine a classroom full of third-grade students, hard at work investigating the effects of friction on the movement of objects. Students are clustered in small groups around the room, taking turns sending furniture sliders down ramps and measuring how far the sliders travel on different types of paper placed at the bottom: printer paper, waxed paper, and sandpaper. The room buzzes with excitement as the students pepper their teacher with observations and questions.

"It went really far on the waxed paper, but it didn't move on the sandpaper."

"I think we messed up on this one. We bumped the ramp, and the slider went farther than the other two times."

"Can I draw a picture to show my data?"

"What would happen if we made the ramp steeper?"

In almost every case, the teacher encourages her students to document these observations and questions in their science notebooks. They take this work seriously, meticulously detailing their findings. When it appears that the groups have finished collecting data, the teacher directs the students to write a claim and support it with evidence from their investigation. The lesson concludes with a class discussion in which students share their claims and respond to each other's thinking.

A look inside the students' notebooks at the end of the lesson reveals highly individualized and meaningful work. Some students have created simple tables to organize their data, while others have listed their measurements in rows. Some have included sketches of the experimental setup and results, while others have written narrative text about their findings. Some entries are brief and lack important information, while others are lengthy and quite detailed. Yet despite these many differences, there is an important commonality. All students were clearly engaged in the *practices* of science: asking questions, analyzing and interpreting data, engaging in argument from evidence, and so on. Their work is a tangible record of their emerging understanding and proficiency—their thinking made visible.

As the teacher reviews her students' work, she not only learns how they are beginning to understand the concept of friction and the crosscutting concept of cause and effect but also ascertains each student's strengths and challenges in participating in the academic work of science. The interdisciplinary nature of this investigation provides students with a meaningful and relevant context for English language arts skills and mathematical practices (incredibly helpful in a crowded elementary curriculum) and provides the teacher with cross-curricular formative assessment. Using notebooks as a source of formative assessment will allow her to plan future lessons to support students in developing science and engineering practices, understanding relevant concepts and vocabulary, and identifying the crosscutting concepts that underlie all scientific content.

The vision set forth in this example demonstrates the power of science notebooks to engage students as active participants in the practice and learning of science. This is the approach I use in my own classroom. As an elementary science specialist, I have successfully used science notebooks with students in first through fifth grades for the past nine years. Though I have made small

Introduction

tweaks and improvements each year in the ways I launch, use, and assess my students' notebooks, the core of my approach has remained the same. I want my students to view their science notebooks as a safe place to write down their thoughts, try out new ideas, pose questions, and work through puzzling data. Although notebook entries serve as a vehicle to help students learn to organize their writing in discipline-specific ways, I also want their entries to reflect their unique voices whenever possible. Above all, I want each student's notebook to be a record of his or her meaningful and personalized learning over the course of our year's work—one that the student may keep and look back on in years to come. These desires influence every aspect of my approach to notebooking, from the model I choose to the ways in which notebooks are used and assessed in my daily lessons.

This book provides what you need to know to adopt a similar approach in your own classroom. I begin by describing an approach to elementary science that aligns with the vision set forth in *A Framework for K–12 Science Education: Practices, Concepts, and Crosscutting Ideas* (NRC 2012) and the *Next Generation Science Standards* (NGSS Lead States 2013), then share research that supports the use of science notebooks in an elementary classroom. Next, I review popular models of science notebooks and explain why I believe that a student-centered approach is the most appropriate for a three-dimensional science classroom. After that, I give details on specific approaches and resources to help you use science notebooks with your students: how to kick off a notebooking practice, ways to help students learn to organize information while also preserving student voice and choice, and lessons and instructional routines that pair well with science notebooks. I discuss how a student-centered approach is a wonderful way to support differentiation, as well as the use of science notebooks in assessment. Finally, I explain my thought processes as I assess student work and plan for future lessons. Additionally, though I recommend having students create their own tables and organizers whenever possible, the appendix includes a sampling of blackline masters for organizational elements (such as a table of contents) and instruction (graphic organizers) for use at your discretion.

As an elementary teacher, I know that every class is unique in its interests, strengths, and challenges—as is every student. It is my hope that you will be inspired by the approach detailed in this book, adopting what works and modifying what doesn't, to implement student-centered notebooks in your own classroom.

References

National Research Council (NRC). 2012. *A framework for K–12 science education: Practices, crosscutting concepts, and core ideas.* Washington, DC: National Academies Press. *https://doi.org/10.17226/13165.*

NGSS Lead States. 2013. *Next Generation Science Standards: For states, by states.* Washington, DC: National Academies Press. *www.nextgenscience.org/next-generation-science-standards.*

CHAPTER 1

A Student-Centered, Three-Dimensional Classroom

The scientist is not a person who gives the right answers, he's one who asks the right questions.

—Claude Lévi-Strauss, *Le Cru et le cuit*

When I was in school, science class consisted of reading from a textbook, taking notes, answering questions, and regurgitating information back to my teachers on tests. There were a few exceptions, most notably a fourth-grade class in which we simulated a space flight to a new planet on an overnight at school; I was a geologist who researched and planned ways to identify the rocks we'd find when we reached our destination. But for the most part, the reasons I fell in love with science were my experiences *outside* the classroom: attending summer workshops at our local science center, participating in hikes and nature programs at our city's metro parks, looking for critters in the creek near our local library, messing around with a microscope and simple chemistry set in my basement. I much preferred *doing* science to *reading about* science (although I loved to read), and fortunately my extracurricular experiences sustained my love for the discipline into college, where I earned degrees in biology and anthropology. But even there, the majority of my experiences were *teacher-centered*, with the professor presenting science as a series of principles and facts. Lab experiences were often sequenced after lecture and typically served as confirmation of the content.

To be fair, I'm sure there were science classrooms across the country that looked dramatically different from the ones in which I sat. More *student-centered* approaches, in which students explore concepts before formalizing their understanding into their own explanations, such as the original three-phase learning cycle, have been in use as early as the 1960s (Atkin and Karplus 1962). Fortunately for today's students, the notion of student-centered science classrooms, in which the emphasis has shifted from the teacher as a purveyor of knowledge to the student as a capable sensemaker, has been much more widely accepted and implemented. We know that it is important for students to roll up their sleeves and engage in the actual work of science: asking questions, conducting investigations, analyzing data, and making sense of it all. Student-centered classrooms are collaborative spaces where students learn from one another as well as the teacher and receive feedback as they hypothesize, investigate, and explore new ideas (Darling-Hammond 1997). Research has supported this shift, finding many benefits to this type of instruction, including increased engagement and a deeper understanding of content (Granger et al. 2012).

Chapter 1

Even with this change, our understanding of best practices continues to deepen, and as a result, our instructional techniques are refined. The most recent and substantial development has been the release of *A Framework for K–12 Science Education*, which lays out a vision for effective student-centered science education: "Students, over multiple years of school, actively engage in scientific and engineering practices and apply crosscutting concepts to deepen their understanding of the core ideas in these fields" (NRC 2012, pp. 8–9). In other words, effective science teaching and learning is *three-dimensional:* an interplay between disciplinary core ideas, science and engineering practices, and crosscutting practices. While many existing resources carefully unpack these elements, a brief consideration of each may be helpful.

Disciplinary core ideas (DCIs) are the *what* of science instruction—the key concepts and principles students must understand to be able to make sense of the natural world. Rather than being isolated facts, DCIs are relevant across multiple branches of science. The *Next Generation Science Standards* (NGSS Lead States 2013) identify disciplinary core ideas in physical science, life science, Earth and space science, and engineering (see Table 1.1). Each DCI contains several related subconcepts, which can be viewed by visiting QR Code 1.1 and clicking on one of the science and engineering domains. DCIs also build on themselves and each other as students become increasingly knowledgeable and capable of abstract thinking. *The NSTA Atlas of the Three Dimensions* (Willard 2020) provides a helpful graphic representation of the flow of disciplinary core ideas across grade bands.

Table 1.1. Disciplinary core ideas of the *Next Generation Science Standards*.

Physical science	Life science	Earth and space science	Engineering, technology, and the application of science
• PS1: Matter and Its Interactions • PS2: Motion and Stability: Forces and Interactions • PS3: Energy • PS4: Waves and Their Applications in Technologies for Information Transfer	• LS1: From Molecules to Organisms: Structures and Processes • LS2: Ecosystems: Interactions, Energy, and Dynamics • LS3: Heredity: Inheritance and Variation of Traits • LS4: Biological Evolution: Unity and Diversity	• ESS1: Earth's Place in the Universe • ESS2: Earth's Systems • ESS3: Earth and Human Activity	• ETS1: Engineering Design

Source: NGSS Lead States (2013).

QR Code 1.1. Disciplinary core ideas learning progressions. Click each core idea to see the subconcepts organized within a particular DCI.

Science and engineering practices (SEPs) are the *how* of teaching and learning—eight practices that "are both a set of skills and a set of knowledge to be internalized" (NGSS Lead States 2013). They represent what students actually do as they investigate and make sense of phenomena (see Figure 1.1). Phenomena are observable events that occur in the universe and that we can use our science knowledge to explain and predict (Achieve, Next Gen Science Storylines, and STEM Teaching Tools 2016). While the SEPs are broad categories that apply to students' work in kindergarten through grade 12, they are broken down into learning progressions that specify what students should be able to do at four grade bands: K–2, 3–5, 6–8, and 9–12. These learning progressions are invaluable for guidance on what a given SEP should look like in your particular classroom. To view each SEP's progression, use QR Code 1.2 and click on a specific science and engineering practice.

Figure 1.1. Science and engineering practices (SEPs) of the *Next Generation Science Standards*.

Science and engineering practices
- Asking questions and defining problems
- Developing and using models
- Planning and carrying out investigations
- Analyzing and interpreting data
- Using mathematics and computational thinking
- Constructing explanations and designing solutions
- Engaging in argument from evidence
- Obtaining, evaluating, and communicating information

Source: NGSS Lead States (2013).

QR Code 1.2. Science and engineering practices learning progressions. Click each practice to view its learning progression across grades K–12.

Conceptualizing eight distinct practices can be overwhelming, and some researchers have proposed grouping them into categories that differentiate their role in science teaching and learning. I particularly like the model proposed by McNeill, Katsh-Singer, and Pelletier (2015), which divides the practices into three categories: investigating, sensemaking, and critiquing. In this model, depicted in Figure 1.2 (p. 4), investigating practices (asking questions, planning and carrying out investigations, and using mathematical and computational thinking) lead to the collection of data about phenomena. Sensemaking practices (developing and using models, analyzing and interpreting data, and constructing explanations) lead to the development of an explanation or model. And critiquing practices (engaging in argument from evidence and obtaining, evaluating, and communicating information) compare this explanation or model with others, with the goal of strengthening or revising the work.

Figure 1.2. Grouping the practices into categories leads to a fluid and flexible model of science.

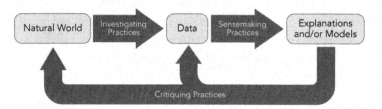

Source: Reprinted with permission from McNeill et al. (2015).

Crosscutting concepts (CCCs) are broad concepts that transcend and connect all scientific disciplines (see Figure 1.3). They don't fit nicely into my 5Ws and an H (who/what/where/when/why/how) analogy for science instruction. Instead, I think of the CCCs as lenses through which students view phenomena and their underlying DCIs. Just as a pair of glasses sharpens and focuses images for their wearer, the CCCs help students make sense of content and concepts across all domains of science. Some researchers and educators take this idea one step further, suggesting that the CCCs have the potential to bridge the gap between "knowing" (the DCIs) and "doing" (the SEPs) when students not only understand how they work but also can use them to make sense of phenomena (Fick, Nordine, and McElhaney 2019). As with the SEPs, the crosscutting concepts are broken down into learning progressions across primary, elementary, middle, and high school grade bands. To view these progressions, use QR Code 1.3 and click on a specific crosscutting concept.

Figure 1.3. Crosscutting concepts (CCCs) of the *Next Generation Science Standards*

Crosscutting concepts
- Patterns
- Cause and effect
- Scale, proportion, and quantity
- Systems and system models
- Energy and matter
- Structure and function
- Stability and change

Source: NGSS Lead States (2013).

QR Code 1.3. Crosscutting concepts learning progressions. Click each concept to view its learning progression across grades K–12.

The key to three-dimensional instruction, as its name suggests, is that the three components—DCIs, SEPs, and CCCs—are interwoven as much as possible. Although you occasionally may need to devote a small amount of time to explicit instruction, particularly in teaching skills encompassed by the SEPs, there should not be a clear division between process and content in a three-dimensional classroom. Rather, a natural interplay occurs among the three: students investigate phenomena (which leads to the development of disciplinary core ideas) using science and engineering practices and keeping the crosscutting concepts in mind as they do. Take, for example, the investigation described in the introduction of this book, in which third-grade students are measuring how far furniture sliders travel across different surfaces when sent down a ramp. They are investigating the disciplinary core idea PS2: Motion and Stability: Forces and Interactions by using several science and engineering practices, including planning and carrying out investigations, analyzing and interpreting data, using mathematical and computational thinking, and constructing explanations and designing solutions. As their teacher leads them in a post-investigation scientist meeting, she will encourage them to reflect on their findings using the crosscutting concept of cause and effect. Here, there is no clear division among the three dimensions. Instead, they overlap and integrate in an engaging and authentic example of science teaching and learning—and of a student-centered science classroom. And all three dimensions are evident in the questions, data collection and analysis, and reflections written in the students' science notebooks.

How can you identify a student-centered, three-dimensional science classroom? The remainder of this chapter discusses some of the many characteristics that provide evidence of such an environment and how they connect to science notebooks.

Valuing and Leveraging Student Experiences

Students arrive at school with wonderfully diverse sets of experience and background knowledge. When it comes to science, this might include exposure to science concepts, time spent in nature, cultural and family conceptions and perceptions of science as a discipline, and sensemaking repertoires. Outstanding student-centered classrooms are places where this diversity of experience is recognized, respected, and leveraged, rather than viewed as a deficit or shortcoming (Bell and Bang 2015).

Active Learning and Engagement

Students in these types of classrooms are not passive consumers of knowledge but actively construct their own learning. What does that look like? It is a misconception that science must be hands-on; active science learning can take many different forms. Hands-on investigation of objects and experimentation are definitely valuable learning experiences, as are working with data from a website or published data set, observing and asking questions about a phenomenon shown in a video or photographs, using simulations to gather data that otherwise could not be collected by hand, and reading and writing nonfiction text. Students ask questions, collect data, and use those data as evidence to construct explanations. Through these varied learning experiences, students gradually construct a deep understanding of scientific principles. Student engagement is typically quite high in these classrooms since they are in charge of their own learning to a great extent. This active learning and engagement translates into high-quality notebook entries, as students take pride in their work.

Teacher as Facilitator

Teachers managing a student-centered classroom stay on the sidelines. Think of a coach or director who constructs routines and strategies but then steps aside and merely gives feedback and guidance while letting the team, ensemble, or cast do the work. The ultimate responsibility for the match or performance does not lie with him or her. In a similar way, teachers in student-centered classrooms act as guides, bringing student attention to phenomena, asking questions to deepen student thinking, and supporting students as they investigate, puzzle, and reason. These teachers view their students as capable partners in this sensemaking journey, and even though they help plan and sequence learning experiences and provide scaffolding when students need support, they resist the temptation to do the work for students or tell them the "correct" answers. Notebook entries from this type of classroom clearly show that students' understanding emerges over time, rather than being perfectly correct responses provided by the teacher.

Productive Talk and Scientific Discourse

A student-centered classroom is not a quiet place or a place where only the teacher's voice is heard. Rather, students do most of the talking—and that talk is productive. Productive talk is talk that promotes learning, and it is an essential feature of a student-centered classroom: "In order to process, make sense of, and learn from their ideas, observations, and experiences, students must talk about them. … Talk forces students to think about and articulate their ideas. Talk can also provide an impetus for students to reflect on what they do—and do not—understand" (NRC 2008, p. 88). Productive talk is collaborative and focused on making students' thinking public, but it is not the default in most science classrooms. Research has shown that most classroom talk falls into an initiation-response-evaluation (I-R-E) model (Cazden 2001). In this model, the teacher poses a question with a "correct" answer (initiation), and students raise their hands and wait to be called on (response). The teacher either affirms a correct response or asks another student until the correct response is given (evaluation). While there is a time and place for I-R-E discourse, such as review of material, it does not elicit student thinking or support sensemaking. Instead, the I-R-E pattern of discourse emphasizes facts and vocabulary terms over the conceptual understanding that underpins three-dimensional learning. Therefore, the I-R-E model is not the dominant model in a student-centered classroom.

Researchers who study classroom discourse focus on the *E* (evaluation) as one place where teachers can dramatically alter the dynamics of their classroom talk. Rather than use this "third turn" in the dialogue (teacher-student-teacher) to affirm or correct student responses, teachers can create space for further conversation with alternative responses (Park et al. 2017). These responses can take many different forms, including questioning strategies and discourse moves (Table 1.2), and talk moves (Table 1.3, p. 8). Note that neither table is a prescriptive list of ways you can encourage students to share their own thinking, deepen their own reasoning, listen to one another, and engage with others' reasoning. These strategies are just some examples of ways skillful teachers help students go public with their thinking (Michaels and O'Connor 2017).

Table 1.2. Teacher discourse moves.

Discourse move	Purpose	Example(s)
Probing	Elicit student observation and ideas	"What did you notice?" "What ideas do you have?"
Pressing	Ask students to reason further or more deeply	"Can you give an example?" "How might you test that idea?" "How does your idea connect with ____?"
Revoicing	Give weight to a student's idea or increase clarity for other students	"So what I hear you saying is ____." "When you said ____, did you mean to say ____?"
Encouraging peer-to-peer talk	Help students engage with each other's ideas directly	"Can anyone add on to what ____ is saying?" "Do you agree with ____? Why or why not?"
Putting an idea on hold	Acknowledge a student's idea while explaining that it won't be addressed at this point	"I like that idea, but let's hold on to that for another time." "Let's come back to that thought later."

Source: Ambitious Science Teaching (2014).

Table 1.3. Talk moves.

Goal: Individual students share, expand, and clarify their own thinking	
Talk move	**Example(s)**
Time to think	Wait time
	Partner talk
	Writing as thinking time
Say more	"Can you say more about that?"
	"Can you give an example of that?"
So, are you saying …?	"What I think you are saying is ___. Did I get that right?"

Goal: Students listen carefully to one another	
Talk move	**Example(s)**
Rephrase or repeat a previous turn	"What ____ said was ___."

Goal: Students deepen their reasoning	
Talk move	**Example(s)**
Ask for evidence or reasoning	"Why do you think that?"
	"What's your evidence?"
Challenge or provide a counterexample	"Does it always work that way?"
	"What if ___?"

Goal: Students think with others	
Talk move	**Example(s)**
Agree/disagree and why	"I agree with ___ because ___."
Add on	"I can add on to what ____ said ___."
Explain what someone else means	"I think I can explain what ____ means ___."

Source: Adapted from Michaels and O'Connor (2012).

A Student-Centered, Three-Dimensional Classroom

In a true student-centered classroom, students are engaged in conversations not only with their teacher but also with their peers. In these classrooms, students share their questions, observations, and emerging ideas with one another. They ask clarifying questions, challenge their peers' thinking, and offer feedback for improvement. Additionally, when you regularly implement talk moves, students may begin to adopt similar responses in their own discussions. Student discourse can happen in pairs, small groups, or even a whole-class setting, and it can be spontaneous or carefully orchestrated by the teacher. No matter the specifics, the underlying premise is the same: The students do a great deal of the talking, and that talk is a foundational component of their learning, which is reflected in their written notebook entries.

Naturally, a community of learners that engages in productive talk is also a community that values listening. This begins with the teacher, who demonstrates respectful, active listening to students as they share their thinking. You can use the discourse and talk moves described in Tables 1.2 and 1.3 in conjunction with neutral and nonjudgmental language to show active listening as you respond to student comments. Rather than affirm or correct a student's thinking, you might reply, "That's an interesting idea. Did anyone else have the same idea? How could we find out if this idea is correct?" Responding in this way keeps the responsibility of thinking with the students and shows that you find their ideas worth listening to.

Students also need strategies and practice to hone their listening skills. I use scientist meetings to introduce this topic with students. We discuss the difference between simply *hearing* someone speak and actually *listening* to someone, and I share my observation that in many classes, students often repeat answers and questions because they are thinking about what they are going to say instead of listening to what others are sharing. In teacher-led discussions, I frequently ask, "Who can repeat what so-and-so said?" and take the time to acknowledge and praise students who demonstrate active listening. Additionally, I use this as an opportunity to introduce and model revoicing or the "So, are you saying …?" talk move, and we discuss how responding in this way demonstrates listening to the speaker. I also encourage students to use these strategies in their small-group and partner discussions.

Classrooms with this kind of productive scientific discourse do not come about by accident. A classroom must have the right climate for productive talk to be successful. You must carefully establish a respectful, equitable climate in which all students feel safe in sharing their thinking with others. You can devote time to teaching students how to ask questions of classmates in respectful ways, so that having a peer ask for further explanation is not perceived as challenging or threatening. Finally, you must send the message that talk is for everyone and not just those deemed to be "smart" or "good at science."

Paying attention to the frequency and duration of student contributions during discussions is important. Some students need to learn to share airspace with their peers, and others need encouragement to contribute to the conversation. Tallying student comments while observing a discussion can be a helpful source of data for reflection on who is actually doing the talking. Other strategies, such as setting a limit on the number of contributions per student in a discussion and having students self-monitor with counters or tally marks, can help students become self-aware of giving others a chance to participate. You can also strategically use wait time and partner talk to support students who might otherwise be anxious about participating in discussion.

Student Choice

Student-centered classrooms necessarily involve some degree of student choice in terms of content, process, or product. In other words, teachers include students in the decision-making processes of planning, teaching, and assessment (Brown 2008). There is a misconception that student choice means an open-ended educational free-for-all, but it can actually take many different forms and be implemented to varying degrees. You may begin the school year by soliciting information about student interests and use this information to frame and shape the investigations, or you might listen carefully to students' questions and initial ideas around a phenomenon and use these to structure the flow of the unit. You can facilitate sensemaking discussions in which students identify what to try next as they dig deeper into a phenomenon. Students can help decide how to demonstrate their learning through varied work products and can help define assessment criteria in the form of a "gotta-have" explanation checklist (Ambitious Science Teaching 2020) or rubric. Additionally, you can gradually increase the amount of student choice over the course of the year as students gain proficiency with concepts, practices, and classroom expectations. Chapter 2 includes one example of an assignment that allows for student choice, and many others are presented in Chapter 6.

Why is student choice important? Research tells us that choice is motivating for students and leads to higher engagement (Evans and Boucher 2015). Allowing students some say in their learning helps build their identities as capable learners and problem solvers and best replicates authentic and lifelong learning experiences.

Using a storyline approach to three-dimensional science learning is a natural way to promote student choice. A storyline, like other approaches to science instruction, is a coherent sequence of lessons, but there is a key difference: Student questions and ideas, not the teacher's, move the storyline forward. As students investigate a phenomenon or attempt to solve a problem, they build their understanding piece by piece, asking questions as they go to refine their thinking. The teacher facilitates this process and assists in assembling the storyline, but the coherence and relevance come from the students (Next Generation Storylines Team 2020).

Visible Student Thinking and Learning

A student-centered classroom is one where students' ideas take center stage, not only in the conversation but also in the physical setting. Many things, from the way student desks or tables are organized to the types of materials displayed on the walls, send powerful messages about whose voices are valued. Some teachers may involve students in the creation of anchor charts or post student work (with permission) as exemplars. Others may regularly share student work with their classes using a document camera or ask students to share their work themselves in a workshop approach.

Instructional strategies also help make student thinking visible. Harvard's Project Zero (2020) has created a number of thinking routines—simple structures that deepen students' thinking as they are repeatedly used over time. These routines can be used across content areas and become increasingly valuable over time as students gain comfort and proficiency with them. (See Chapter 6 for a discussion of how these routines pair beautifully with science notebook entries.)

You can also employ a strategy known as documentation to help bring student learning to the forefront of your classroom. A key component of the Reggio Emilia approach to education, documentation simply involves capturing student ideas and creating visible representations of student learning (Project Zero 2006). Documentation can be used for several different purposes and audiences: to help you reflect on student progress, to spark student reflection and deeper learning, and to share student learning with a broader audience. Regardless of the purpose, student's age, or content area, documentation of student thinking and learning centers student voice and agency in the classroom.

Emphasis on Conceptual Understanding and Transfer Learning

Unlike my experience as a science student, instruction in a student-centered, three-dimensional classroom is not focused on memorizing isolated facts and terminology. Instead, significant attention is placed on conceptual understanding, and students learn vocabulary only as needed to bolster and extend that understanding. The *Framework* and *NGSS* support this emphasis with the disciplinary core ideas and crosscutting concepts. As students investigate and make sense of phenomena, they are building a highly personalized and meaningful understanding of broad science concepts that often transcend the typical delineations among scientific disciplines. Unlike memorized facts, which are quickly forgotten, a conceptual understanding built on the active learning strategies present in a student-centered, three-dimensional classroom is more likely to stick with students even after they've moved on to a new investigation (Minner, Levy, and Century 2010).

Students who have developed a robust conceptual understanding of scientific principles are better able to apply their knowledge to novel situations. Teachers in these types of classrooms use instructional strategies that promote transfer learning and actively seek out opportunities for students to apply what they've learned in new contexts. For example, metacognitive strategies, a foundational component of student-centered instruction, may improve the durability and transfer of learning (Georghiades 2000).

Interdisciplinary, Cross-Curricular Work

No silos here! Critics of our traditional education system argue that authentic learning is not segmented into discrete disciplines (Honebein, Duffy, and Fishman 1993), and a student-centered, three-dimensional classroom represents a major step toward addressing that concern. When students are engaged in meaningful and authentic investigations, other subjects such as art, math, and literacy (which encompasses reading, writing, listening, speaking, and viewing) are naturally incorporated as students make observations, collect data, read to confirm or extend their findings, and communicate what they have learned. Having an authentic context in which to practice academic skills and strategies is motivating for students and can lead to deeper, more thoughtful work.

Student-centered classrooms are ones in which students' thinking and learning take center stage. Even though they are highly individualized, reflecting their particular student populations, these classrooms have many elements in common. The next chapter examines how science notebooks are a natural component of these active and exciting learning spaces.

References

Achieve, Next Gen Science Storylines, and STEM Teaching Tools. 2016. Using phenomena in *NGSS-designed lessons and units.* STEM Teaching Tools Initiative, Institute for Science + Math Education. Seattle: University of Washington. *http://stemteachingtools.org/brief/42.*

Ambitious Science Teaching. 2014. A discourse primer for science teachers. Ambitious Science Teaching Development Group. *https://ambitiousscienceteaching.org/wp-content/uploads/2014/09/Discourse-Primer.pdf.*

Ambitious Science Teaching. 2020. "Gotta-have" explanation checklist. Ambitious Science Teaching Development Group. *https://ambitiousscienceteaching.org/tools-face-to-face/#Gottahave.*

Atkin, J. M., and R. Karplus. 1962. Discovery or invention? *The Science Teacher* 29: 45–47.

Bell, P., and M. Bang. 2015. How can we promote equity in science education? Practice Brief 15. STEM Teaching Tools. *http://stemteachingtools.org/brief/15.*

Brown, J. K. 2008. Student-centered instruction: Involving students in their own education. *Music Educators Journal* 94 (5): 30–35.

Cazden, C. B. 2001. *Classroom discourse: The language of teaching and learning.* 2nd ed. Portsmouth, NH: Heinemann.

Darling-Hammond, L. 1997. *The right to learn: A blueprint for creating schools that work.* San Francisco: Jossey-Bass.

Evans, M., and A. R. Boucher. 2015. Optimizing the power of choice: Supporting student autonomy to foster motivation and engagement in learning. *Mind, Brain, and Education* 9 (2): 87–91.

Fick, S. J., J. Nordine, and K. W. McElhaney, eds. 2019. *Proceedings of the summit for examining the potential for crosscutting concepts to support three-dimensional learning.* Charlottesville: University of Virginia. *http://curry.virginia.edu/CCC-Summit.*

Georghiades, P. 2000. Beyond conceptual change learning in science education: Focusing on transfer, durability and metacognition. *Educational Research* 42 (2): 119–139.

Granger, E. M., T. H. Bevis, Y. Saka, S. A. Southerland, V. Sampson, and R. L. Tate. 2012. The efficacy of student-centered instruction in supporting science learning. *Science* 338 (6103): 105–108.

Honebein, P. C., T. M. Duffy, and B. J. Fishman. 1993. Constructivism and the design of learning environments: Context and authentic activities for learning. In *Designing environments for constructive learning,* ed. T. M. Duffy, J. Lowyck, and D. H. Jonassen, 87–108. Berlin: Springer.

Lévi-Strauss, C. 1964. *Le Cru et le cuit.* Paris, France: Plon.

McNeill, K., R. Katsh-Singer, and P. Pelletier. 2015. Assessing science practices: Moving your class along a continuum. *Science Scope* 39 (4): 21–28.

Michaels, S., and C. O'Connor. 2012. *Talk science primer.* Cambridge, MA: TERC. *https://inquiryproject.terc.edu/shared/pd/TalkScience_Primer.pdf.*

Michaels, S., and C. O'Connor. 2017. From recitation to reasoning: Supporting scientific and engineering practices through talk. In *Helping students make sense of the world using Next Generation Science and Engineering Practices*, ed. C. V. Schwarz, C. Passmore, and B. J. Reiser, 311–336. Arlington, VA: NSTA Press.

Minner, D. D., A. J. Levy, and J. Century. 2010. Inquiry-based science instruction—what is it and does it matter? Results from a research synthesis years 1984 to 2002. *Journal of Research in Science Teaching* 47 (4): 474–496.

National Research Council (NRC). 2008. *Ready, set, SCIENCE! Putting research to work in K–8 science classrooms.* Washington, DC: National Academies Press. *https://doi.org/10.17226/11882.*

National Research Council (NRC). 2012. *A framework for K–12 science education: Practices, crosscutting concepts, and core ideas.* Washington, DC: National Academies Press. *https://doi.org/10.17226/13165.*

Next Generation Storylines Team. 2020. What are storylines? *www.nextgenstorylines.org/ what-are-storylines.*

NGSS Lead States. 2013. *Next Generation Science Standards: For states, by states.* Washington, DC: National Academies Press. *www.nextgenscience.org/next-generation-science-standards.*

Park, J., S. Michaels, R. Affolter, and C. O'Connor. 2017. Traditions, research, and practice supporting academically productive classroom discourse. *Oxford Research Encyclopedia of Education. https://doi. org/10.1093/acrefore/9780190264093.013.21.*

Project Zero. 2006. Making learning visible: Understanding, documenting, and supporting individual and group learning. Harvard Graduate School of Education. *http://mlvpz.org/index.html.*

Project Zero. 2020. Visible thinking. Harvard Graduate School of Education. *https://pz.harvard.edu/ projects/visible-thinking.*

Willard, T. 2020. *The NSTA atlas of the three dimensions.* Arlington, VA: NSTA Press.

CHAPTER 2

Science Notebooks as a Learning Tool

An experiment is a question which science poses to Nature, and a
measurement is the recording of Nature's answer.

—Max Planck, *Scientific Autobiography and Other Papers*

A science notebook is an important tool that supports three-dimensional instruction by providing space for students to pose questions, grapple with data, and think through various interpretations of those data as they determine what counts as evidence to support answers to the questions. Additionally, there are many other compelling reasons for the regular use of a science notebook in an elementary classroom.

Notebooks Allow Students to Mirror the Practice of Actual Scientists

Recording investigations in a notebook is not simply an activity created by science educators—it has been an integral part of the everyday work of scientists throughout history and around the world. Bringing the practice into the elementary classroom adds authenticity to the instructional activities in which students are engaged. This can be easily enhanced by having students view and discuss pages from historical scientific notebooks, an activity made easier with the increasing number of notebooks that have been digitized and made available online. With explicit connections and discussion, students can view themselves as part of the larger scientific enterprise. (See Chapter 4 for more strategies to bridge the classroom and the lab or field.)

Additionally, connecting students' science notebooks to the work of scientists can help students develop an understanding of the nature of science, an essential understanding for students in the 21st century. As highlighted in *A Framework for K–12 Science Education*, "There is a strong consensus about characteristics of the scientific enterprise that should be understood by an educated citizen" (NRC 2012, p. 78). The *Next Generation Science Standards* take this idea one step further, devoting an entire appendix to the nature of science (NGSS Lead States 2013, Appendix H). In this appendix, the authors identify eight basic understandings about the nature of science:

- Scientific investigations use a variety of methods.
- Scientific knowledge is based on empirical evidence.

- Scientific knowledge is open to revision in light of new evidence.
- Scientific models, laws, mechanisms, and theories explain natural phenomena.
- Science is a way of knowing.
- Scientific knowledge assumes an order and consistency in natural systems.
- Science is a human endeavor.
- Science addresses questions about the natural and material world.

Although the second-to-last understanding, "Science is a human endeavor," is the most closely aligned to the use of notebooks in science classrooms, the other seven can be easily brought to light with purposeful conversation around the types of activities conducted and recorded in student notebooks. For example, providing students with opportunities to revisit and revise thinking in their notebooks based on new evidence *and* engaging in a brief discussion about how this is similar to what scientists do in their everyday work build the important understanding that scientific knowledge is not fixed or static. Additionally, this type of exercise promotes metacognition and helps students identify successes and growth in their learning, which can be a powerful form of motivation.

Notebooks Are Sensemaking Tools

Science's goal is to explain natural phenomena, and the same is true for science education. Students arrive at school with a multitude of experiences with the natural and material world, and high-quality science instruction leverages those experiences as students ask questions, collect evidence, and develop explanations. This is especially true with the phenomenon-based, three-dimensional approach to science education presented in the *Framework* and *NGSS*. Studies have shown that journaling supports a classroom environment that allows students to make sense of complex science topics (Audet, Hickman, and Dobrynina 1996).

Sensemaking is a messy and iterative business, requiring students to test out ideas, think through data, and synthesize information from multiple sources. This process happens over and over with the collection of new evidence, consideration of a new perspective, or response to a question posed by a teacher or peer. Science notebooks provide a safe space for students to work through these steps without worrying about a grade or having the "correct" answer. In Chapter 8, I discuss how I use notebooks for formative assessment and why I prefer to have students share their final thinking about a concept or phenomenon (summative assessment) outside their notebooks whenever possible.

Notebooks Are a Record of Learning and Growth

A science notebook serves as an evolving record of a student's learning over the course of a unit and a year, including not just what was investigated and taught but also the highly personalized account of how that particular student made sense of the phenomena and underlying concepts. Making thinking visible through formal and informal writing helps students articulate their understanding and identify what they are still working to understand. In this way, a science notebook is both a thinking tool for the student and a rich source of data for formative assessment by the teacher (Aschbacher and Alonzo 2006).

This ever-evolving, real-time document affords a natural opportunity for reflection. In my experience, students enjoy revisiting prior work and naturally comment on different aspects of their academic growth, from the quality of their handwriting to the detail in their sketches to their facility with data analysis and argumentation. You can capitalize on this natural proclivity for reflection, inviting students to review older entries and discuss or write about how their understanding has changed and how their proficiency with science and engineering practices has grown over time. (See Chapter 6 for some strategies that have helped my students deepen their reflections on their work.)

Notebooks Allow for Voice and Choice

Who are your favorite authors? Why do you enjoy reading their work? Chances are that it has something to do with their writing style or the way they express ideas. Literacy educators refer to this as the author's voice, and it is a critical component of effective written communication. Researcher Donald Graves explains *voice* as "the imprint of ourselves on our writing" (1983, p. 227), and it applies to all types of writing, including the technical writing found in science notebooks. Having read thousands of science notebook entries over the course of my career, I can attest that even when students are given a teacher-directed prompt, their prior experiences, schemas, and personalities lead to unique entries. For example, consider the two entries shown in Figure 2.1, which were written by students observing the enlarged pistils of their Wisconsin Fast Plants by shining a flashlight on them to visualize the seeds inside. Both students wrote about the same observations, but each student's voice is quite apparent: One adopts a formal, scientific tone (2.1a), and the other writes their observations in the same way they would explain them to a peer (2.1b).

Figure 2.1. Students' unique voices come through in their notebook entries.

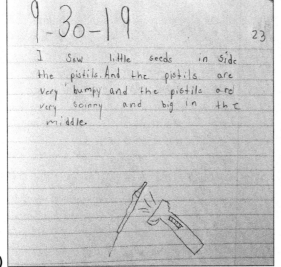

(a)

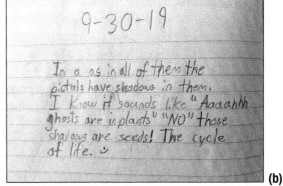

(b)

One student uses a formal, scientific tone to make observations (a), and another student writes as if they were talking to a peer (b).

Voice is crafted through practice and experience. Students need time to write, share, receive feedback from teachers and peers, and revise. Tom Romano, a researcher and professor, notes that teachers "must create the kind of classroom atmosphere where students feel free to express themselves—where both error and accomplishment are natural, expected parts of learning" (2004, p. 21). Note that these characteristics are quite similar to the vision of a three-dimensional classroom set forth in Chapter 1. By engaging in three-dimensional learning experiences and taking academic risks, students can build their understanding of science concepts and develop their voices as technical writers.

A second important opportunity presented by science notebooks is choice. You can capitalize on the highly individualized nature of sensemaking and learning by giving students opportunities to make choices about how they represent and organize content from their investigations. Providing choice in assignments can increase student creativity, agency, and motivation (Kohn 1993; Evans and Boucher 2015).

For example, consider the three notebook entries shown in Figure 2.2. These three third-grade students were engaged in the same investigation: observing the different life cycle stages of a mealworm (larvae, pupae, and adult) and recording their observations in both sketches and descriptive writing. As the students were given only this broad instruction, they were free to choose an organizational style and format that made sense to them, and their final products were quite different. One student opted to focus on the details of one stage at a time, creating an enlarged sketch surrounded by descriptions (2.2a). (She did the same for the other life cycle stages on subsequent pages.) Another student blended observation and imagination as she added a face and fictional details to her sketch of the mealworm larvae (2.2b). A third student took a comparative approach, organizing descriptive details into a table that showed all three life cycle stages at a glance (2.2c). In Chapter 9, I revisit these three student work samples and discuss how student choices in this assignment led me to personalize feedback and guidance for each student in future activities.

Although choice is a powerful tool for motivation and engagement, not every notebook entry lends itself to student choice. Science teachers are tasked with guiding their students toward proficiency in the practices of science, including documenting observations, working with data, and communicating findings in writing. While students can certainly achieve proficiency through the trial and error that choice provides (along with consistent teacher feedback), this can take a great deal of time. At times, it may be preferable to have students work within strict guidelines or even use teacher-created organizers in their notebook entries to help them practice and eventually master a specific strategy or practice. (See Chapter 5 for a gradual release of responsibility model that can be applied in these situations.) Nonetheless, I strongly recommend that you consider when and how to allow for greater student freedom and choice in their notebook entries.

Notebooks Allow You to Meet Students Where They Are

Our classrooms are filled with diverse learners with unique strengths and areas of needed growth. Meeting the needs of these students while providing science instruction can be a great challenge. When you make space for student voice and choice in notebooks, you are embedding an authentic form of differentiation into your classroom as you allow students to engage with the content in a way that makes sense to them.

Figure 2.2. Students' different choices of style and format in recording observations of mealworm life cycle stages.

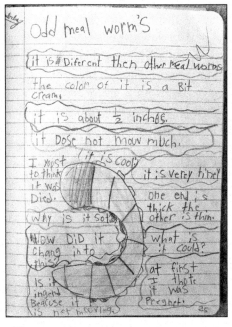

(a)

(b)

(c)

One student focused on the details of one stage at a time (a); another student blended observation and imagination (b); and yet another organized descriptive details into a table (c).

Chapter 2

Reviewing students' notebook entries gives me great insight into not only their understanding of the content but also their proficiency with writing and science practices. Providing verbal and written feedback on notebook entries allows me to personalize my instruction and help move all students forward, no matter their starting point. It also means that I can always find something positive in a student's work to praise alongside my suggestions for improvement.

Consider again the three mealworm observations in Figure 2.2. These three students are in the same grade but clearly have varying levels of proficiency in documenting their observations in a scientific manner. These differences are clearly shown only because of the open-ended nature of the assignment and the fact that I allowed students to choose how to represent and organize their observations. If I had provided a graphic organizer with a box for a sketch and sentence starters or blanks to fill in, I would not necessarily have learned that student two needed to talk about the differences between creative drawings and observational sketches and that student three clearly is ready for extension in this area. I further explain my observations and insights into this student work, along with my instructional decisions, in Chapter 9.

Notebooks Develop Literacy Skills

A complete understanding of literacy involves not just reading and writing but also listening, speaking, and viewing (Van Kraayenoord 2001). While these are obviously components of regular literacy instruction in elementary classrooms, they also have far-reaching cross-curricular overlap. Research has demonstrated that situating these practices within disciplinary content such as science is an effective way to improve student proficiency (Cervetti et al. 2012). A thoughtfully planned and explicitly implemented science notebook is an important tool in all facets of students' literacy development.

Writing is the most obvious connection between science notebooks and literacy development. Science notebooks afford the opportunity for a variety of types of expository and informational writing: note-taking, predictions, reflections, and argumentation, to name a few. Writing can be used at the beginning of an investigation to generate questions and identify preliminary ideas, during an investigation to document procedures and record observations, and after an investigation as a means of sensemaking and identifying new questions for study. An advantage to writing in this context is that students' firsthand experience through investigations and simulations gives them a depth of knowledge from which they can draw, resulting in higher quality writing.

When you provide opportunities for students to discuss their notebook entries with each other, you also provide authentic opportunities for speaking and listening. Productive talk can be used before writing to help students generate ideas and encourage reluctant writers to get something on the page, or it can be used after writing so students may consider alternative perspectives and interpretations of data. Students can also give and receive feedback on peers' notebook entries with the goal of revision. Finally, students can share their notebook entries and strategies in a type of writing workshop so that others can emulate them, a process described in more detail in Chapter 5.

As discussed in Chapter 1, these types of conversation are essential for students engaged in the process of sensemaking, but implementing and managing this discourse can be difficult for teachers (Harris, Phillips, and Penuel 2012). To maximize the benefit of productive talk, you should establish norms for discussion and teach, practice, and reteach student roles and language (Bacolor et al. 2020). Both the teacher discourse moves (Ambitious Science Teaching 2014) and the student talk moves (Michaels and O'Connor 2012) discussed in Chapter 1 are helpful tools for doing so.

Science notebooks also pair well with reading and viewing. Students can use their notebooks to take notes from or respond to informational text or to write about how the information from their reading supports and extends what they have found in their investigations. Thinking routines such as See/Think/Wonder (discussed in more detail in Chapter 6) provide a lens for viewing phenomena. Additionally, interpreting data tables and different types of graphs can also be considered discipline-specific forms of viewing and reading. A science notebook is thus a perfect companion for activities that are traditionally viewed as part of a literacy block, making it possible for you to authentically integrate your science and literacy instruction into inquiries and investigations that transcend traditional disciplinary labels.

While vocabulary development is not a specific process, it underpins several components of literacy and thus deserves to be addressed as well. Science is a vocabulary-heavy discipline, and as Rupley and Slough explain, "The specialized vocabulary knowledge in science represents the concept-laden hooks on which learning is hung and enables students to build prior knowledge through the expansion of these conceptual hooks" (2010, p. 100). Rupley and Slough articulate an essential truth regarding science vocabulary: Vocabulary terms represent concepts. It is crucial, therefore, that students demonstrate conceptual understanding *before* being expected to proficiently use correct terminology for those concepts. This is exactly the opposite of the instructional approach in many classrooms, in which vocabulary is front-loaded before active investigation, and students are expected to learn definitions without first understanding the underlying principles. Science educators who advocate for reversing this traditional front-loaded instruction use a handy mnemonic: ABC-CBV, which stands for "activity before concept; concept before vocabulary." As simply described in this mnemonic, science instruction should allow students to engage and investigate for themselves before being introduced to the formal scientific concepts, and to demonstrate an understanding of the concepts before being introduced to scientific vocabulary.

It is important to be aware that even though a science notebook is a great tool for practicing literacy skills and strategies, literacy plays an essential role in science teaching and learning. Reading, writing, listening, speaking, and viewing are not a separate discipline to be integrated into science, but rather an integral part of how students engage with science. P. David Pearson reminds us that "reading and writing are always better when they are tools not goals. ... Reading and writing are not about reading and writing in general, but about reading and writing particular texts that are grounded in particular experiences" (2006). Hand and Choi similarly point out that "writing is a means (a tool) for helping students engage with and construct understanding of science. It is not simply a tool that students have to learn to use, but rather is a tool that by its very use will help students learn" (2009, p. 293). Though these researchers specifically focus on reading and writing in their work, the same case can be made for the other components of literacy. You may recognize and appreciate the benefit from the overlap between the two disciplines, yet you should still strive to ensure that the literacy-related work you ask students to do in their science notebooks is authentically connected to your three-dimensional instruction and not just practice for practice's sake.

Notebooks Can Raise Student Achievement

If the previous six reasons weren't enough to persuade you to implement science notebooks in your classroom, perhaps this final one will: Notebooks can have a positive impact on student achievement. Studies have shown that the use of academic notebooks, including science notebooks, leads to growth in students' abilities to use academic vocabulary to communicate their understanding and achievement (Huerta et al. 2016; Huerta and Spies 2016; Marzano 2004; Rheingold, LeClair, and Seaman 2013). By supporting and complementing the three-dimensional instruction in an elementary classroom, science notebooks are a powerful tool for student learning and growth.

In this chapter, I've shared many reasons to consider using science notebooks as a learning tool in your classroom. Notebooks allow students to authentically engage in work similar to that of actual scientists, but equally important, they provide opportunities for student choice, allow for differentiation, and serve as valuable sources of assessment. Notebooks are perfectly suited to develop literacy skills and can improve student achievement. In the next chapter, I review several different notebook models, discuss important ways in which the work done by students in science classrooms is different from that of scientists, and examine how these differences influence the selection of an appropriate notebook model for elementary students.

References

Ambitious Science Teaching. 2014. A discourse primer for science teachers. Ambitious Science Teaching Development Group. *https://ambitiousscienceteaching.org/wp-content/uploads/2014/09/Discourse-Primer.pdf.*

Aschbacher, P., and A. Alonzo. 2006. Examining the utility of elementary science notebooks for formative assessment purposes, *Educational Assessment* 11 (3–4): 179–203. *https://doi.org/10.1080/10627197.2006.9652989.*

Audet, R. H., P. Hickman, and G. Dobrynina. 1996. Learning logs: A classroom practice for enhancing scientific sense making. *Journal of Research in Science Teaching* (33): 205–222. *https://doi.org/10.1002/(SICI)1098-2736(199602)33:2<205::AID-TEA5>3.0.CO;2-Y.*

Bacolor, R., T. Cook-Endres, T. Lee, and A. Allen. 2020. How can I get my students to learn science by productively talking with each other? STEM Teaching Tools. *http://stemteachingtools.org/brief/6.*

Cervetti, G. N., J. Barber, R. Dorph, P. D. Pearson, and P. G. Goldschmidt. 2012. The impact of an integrated approach to science and literacy in elementary school classrooms. *Journal of Research in Science Teaching* 49 (5): 631–658.

Evans, M., and A. R. Boucher. 2015. Optimizing the power of choice: Supporting student autonomy to foster motivation and engagement in learning. *Mind, Brain, and Education* 9 (2): 87–91.

Graves, D. 1983. *Writing: Teachers and children at work.* Portsmouth, NH: Heinemann.

Hand, B., and A. Choi. 2009. Writing in classroom science. In *The world of science education: Handbook of research in North America,* vol. 2/1, ed. W.-M. Roth and K. Tobin, 293–305. Rotterdam, Netherlands: Brill Sense.

Harris, C. J., R. S. Phillips, and W. R. Penuel. 2012. Examining teachers' instructional moves aimed at developing students' ideas and questions in learner-centered science classrooms. *Journal of Science Teacher Education* 23 (7): 769–788.

Huerta, M., and T. G. Spies. 2016. Science inquiry and writing for ELLs: A gateway for building understanding and academic language. *Science Activities* 53 (1): 24–32.

Huerta, M., F. Tong, B. J. Irby, and R. Lara-Alecio. 2016. Measuring and comparing academic language development and conceptual understanding via science notebooks. *Journal of Educational Research* 109 (5): 503–517. *https://doi.org/10.1080/00220671.2014.992582.*

Kohn, A. 1993. Choices for children: Why and how to let students decide. *Phi Delta Kappan* 75 (1): 8–20. *www.jstor.org/stable/20405017.*

Marzano, R. J. 2004. *Building background knowledge for academic achievement.* Alexandria, VA: ASCD.

Michaels, S., and C. O'Connor. 2012. *Talk science primer.* Cambridge, MA: TERC. *https://inquiryproject.terc. edu/shared/pd/TalkScience_Primer.pdf.*

National Research Council (NRC). 2012. *A framework for K–12 science education: Practices, crosscutting concepts, and core ideas.* Washington, DC: National Academies Press. *https://doi.org/10.17226/13165.*

NGSS Lead States. 2013. *Next Generation Science Standards: For states, by states.* Washington, DC: National Academies Press. *www.nextgenscience.org/next-generation-science-standards.*

Pearson, P. D. 2006. "Letter to the Editor." *New York Times,* March 28.

Planck, M. 1949. *Scientific autobiography and other papers.* New York: Philosophical Library.

Rheingold, A., C. LeClair, and J. Seaman. 2013. Using academic notebooks to support achievement and promote positive classroom environments. *Middle School Journal* 45 (1): 24–32.

Romano, T. 2004. The power of voice. *Educational Leadership* 62 (2): 20–23. *www.ascd.org/el/articles/ the-power-of-voice.*

Rupley, W. H., and S. Slough. 2010. Building prior knowledge and vocabulary in science in the intermediate grades: Creating hooks for learning. *Literacy Research and Instruction* 49 (2): 99–112.

Van Kraayenoord, C. 2001. Literacy for all: Reading, writing, speaking, listening, and viewing. *International Journal of Disability, Development and Education* 48 (4): 327–329.

CHAPTER 3

Choosing the Right Notebook Model

The only difference between screwing around and science is writing it down.

—Alex Jason, *Mythbusters*

Elementary science notebooks are enjoying a surge in popularity, with many resources online and in print. Even a quick web search shows many different approaches and models for notebooks, but not all are appropriate for the developmental needs of elementary students or the vision of three-dimensional learning discussed in Chapter 1. In this chapter, I review several popular models of notebooks and discuss the approach that I've found to be the most effective.

It is important to keep in mind differences between *science as a discipline* (what is practiced by researchers) and *school science* (what is practiced by students). To be sure, there are many significant overlaps between the two. In effective three-dimensional science classrooms, students engage in many of the same behaviors as scientists: collecting and analyzing data, reasoning about evidence, constructing and refining models, working cooperatively, sharing information orally and in writing, and so on. In fact, teachers' ultimate goal as science educators is to help apprentice students into the cognitive discipline of science (Collins, Brown, and Newman 1989; Collins, Brown, and Holum 1991). Yet there are key differences that should be considered, particularly in terms of the background knowledge that each group brings to the table. Scientists have done research, made evidence-based claims, extensively read the literature, and are by-products of extensive formal schooling (K–12 and higher education). These experiences contribute to a rich background knowledge that informs their methods, approaches, and ways of thinking about data. While students also have background knowledge, it is mostly from their lived experiences and prior classroom experiences and is thus limited in scope compared with that of scientists.

Although elementary students come to school with a wealth of knowledge about the natural world and are capable of sophisticated reasoning (NRC 2007), they still need carefully planned learning experiences to deepen their content knowledge and appropriately label science concepts with discipline-specific vocabulary. An elementary student's science notebook thus might include graphic organizers such as Frayer models to assess whether students have ownership of the concept behind the term or occasional space for taking notes on or responding to nonfiction text, which are not typically found in researchers' notebooks. Additionally, elementary students are

Chapter 3

developing proficiency in academic skills (such as reading and writing), executive functioning skills (such as organization), and discipline-specific skills (such as measurement and argumentation). Effective approaches to instruction in general, and notebooks in particular, provide support and scaffolding in these areas.

Any effective science notebook model should reflect the integrated, multifaceted approach to science education that is three-dimensional learning. This approach does not separate content and process, but instead weaves them together throughout instruction. These two lenses—using a three-dimensional approach to learning and meeting the specific needs of elementary students—aid in examining several popular models of notebooks: lab notebook, field journal, and interactive notebook, summarized in Table 3.1.

Lab Notebook

A lab notebook is a formal record of experiments performed and data collected. This type of notebook is quite formal and structured. Entries are dated and typically contain a hypothesis or purpose (or both), materials and experimental procedure, raw data, and some analysis of the data. The emphasis is on the documentation of the exact procedure followed, so others may replicate the experiment, and the collection of data.

In a research lab, the notebook is a legal document. Strict guidelines are thus in place: Bound notebooks are often used so that pages cannot be removed without notice, and entries are written in ink so that they may not be erased. While lab notebooks used in classrooms are not official documents in the same way, some of these guidelines may be implemented by teachers who had personal experience with keeping lab notebooks in college science courses.

Alignment With Three-Dimensional Learning

Lab notebooks, as a record of activity and findings, emphasize process over content and are thus most closely aligned with a single dimension (i.e, SEPs) instead of the interwoven emphasis on science and engineering practices, disciplinary core ideas, and crosscutting concepts. And while the lab notebook certainly includes a heavy emphasis on certain SEPs (planning and carrying out investigations, analyzing and interpreting data, using mathematics and computational thinking), others are less emphasized or absent altogether (developing and using models; engaging in argument from evidence; obtaining, evaluating, and communicating information). Furthermore, lab notebooks epitomize the stereotypical view of science as a discipline: a single scientist working alone in a lab and following a rigid procedure of doing science (often referred to as the "scientific method"). Some researchers do follow this approach, but studies have shown that "there is no single 'scientific method' universally employed by all" (NRC 2010, p. 19) and that scientists use a much more diverse set of approaches in their work. Additionally, the science education community recognizes that science is a social enterprise, unlike the solo record of work kept in this type of notebook.

Table 3.1. Comparison of popular notebook models.

	Primary use	Characteristics	Alignment with 3D learning	Benefits for elementary students	Drawbacks for elementary students

Lab notebook	Formal record of experiments and data in research labs	• Dated entries
		• Purpose and/or hypothesis
		• Materials
		• Experimental procedure
		• Raw data
		• Data analysis
			• Some SEPs emphasized:
			  * Planning and carrying out investigations
			  * Analyzing and interpreting data
			  * Using mathematics and computational thinking
			• DCIs and CCCs not an explicit focus
				• Organized and easy to read
				• Predictable structure
				• Opportunity to develop strong data collection and analysis skills
				• Alignment with mathematical practices
					• Some SEPs ignored
					• Emphasizes process over content
					• Emphasizes stereotypical view of science
					• Little opportunity for sensemaking
					• No opportunities for note-taking or practice using vocabulary
					• Reflects solo work

Field journal	Used in field to describe conditions, observations, and collections	• Dated entries • Time, weather, and other salient conditions • Quick notes or coding system that may be transcribed later • Sketches	• Can provide a basis for documenting and describing phenomena • Real-world context and data	• Integrates sketching and description • Engages students who prefer artistic forms of expression	• Emphasizes documentation over sensemaking • No opportunities for note-taking or practice using new vocabulary • Reflects solo work • Better suited to some science disciplines than others
Interactive notebook	Used in classrooms to aid in acquisition of science content knowledge	• Two-page spreads: input and output • Notes • Graphic organizers • Foldables	• Addresses disciplinary core ideas	• Engages students with content • Adds creative and interactive elements • Includes elementary-specific tasks	• Emphasizes content over practices • Structure can be limiting • Emphasizes teacher as content provider • Can lead to checklist assessment

Chapter 3

Meeting the Needs of Elementary Students

Lab notebook entries are highly structured and predictable, two characteristics that can help build students' confidence in their abilities to plan and conduct investigations and collect and analyze data. The structure helps students organize their thinking and makes entries easier for both students and teachers to read. The emphasis on data collection and analysis strongly aligns with mathematical practices and standards, such as the measurement and data domain of the *Common Core State Standards for Mathematics* (NGAC and CCSSO 2010) and can help students develop these important skills. For these reasons, elementary teachers may find utility in using this model with their students as appropriate.

Despite these benefits, this model is not sufficient as a single approach for elementary classrooms. It does not provide the support and scaffolding necessary for the cognitive needs of young students, and it emphasizes the documentation of process over sensemaking and knowledge construction. Moreover, it affords no opportunities for the range of other instructional activities used in elementary science classrooms. Finally, the rigid structure may become a challenge for students who struggle with writing, leading to a loss of confidence and interest in the subject.

Field Journal

As the name implies, a field journal is taken into the field to document what is observed or collected. Entries are dated and include the physical location, the weather, other salient environmental conditions, and observations. These might include descriptions of species encountered, behavior of animals, and any specimens collected. Some advocate taking quick notes or using a coding system to document animal behavior and then transcribing these notes into more formal writing later on. Some field journals also include sketches of organisms observed or the locations of objects and organisms.

Alignment With Three-Dimensional Learning

The observation and explanation of phenomena is a key component of three-dimensional instruction (see Chapter 1). Field journals can serve as vehicles for documenting observations of a phenomenon that students may return to throughout a lesson or over the course of an entire unit. Field journal entries can also provide real-world qualitative and quantitative data that students can use to construct explanations of a phenomenon.

However, just as with lab notebooks, there are limits to the degree to which field journals align with three-dimensional instruction. Many of the science and engineering practices do not naturally emerge from the use of a field journal, and the focus is on documentation, not sensemaking. This means that like lab notebooks, field journals are not the best model for the understanding of disciplinary core ideas and crosscutting practices.

Meeting the Needs of Elementary Students

There is great value in teaching elementary students to be careful observers of the natural world, and you will see many field journal–inspired pages in my approach to notebooks throughout the remainder of this book. I've found that students who observe closely are more likely to generate thoughtful questions that lead to rich investigations. Field journals also provide opportunities for interdisciplinary learning, including art and descriptive writing. Additionally, this type of documentation is accessible and nonthreatening. All students, regardless of language or writing proficiency, can make their own observations and interpretations that can be celebrated and refined.

As with the lab notebook model, the field journal cannot be the only model used with elementary students, for a variety of reasons. The range of disciplinary content studied by elementary students exceeds what is typically found in a field journal; physical science concepts, for example, don't mesh well with this approach. Furthermore, the purpose of a field journal is often primarily documentation, and any approach used with students must include opportunities for sensemaking, practice using newly acquired vocabulary, argumentation, and so on. Finally, field journals tend to be the product of solo work and don't reflect the communication and exchange of ideas that occur in elementary science classrooms. Despite these limitations, you can selectively use this model at appropriate times within a broader approach to notebooks.

Interactive Notebook

Unlike lab notebooks and field journals, an interactive notebook is a model specifically created for students and school science. This model can be described as an "organizational system for class notes … and a creative outlet for students as they process information from class" (Robinson 2018, p. 20). An interactive notebook blends teacher-directed information (input) and student-generated responses (output). Typically, this information is organized in a two-page spread, with input on the right-hand page and output on the left, or vice versa. Input may take the form of notes from a lecture or video, vocabulary, a graphic organizer, or a foldable with science content. The output pages are for students to interact with the content through brainstorming, written reflections, drawings, diagrams, or other higher-order thinking activity (Chesbro 2006). Proponents of this model explain that it aids in the learning of scientific content through increased participation and processing (Young 2003).

Alignment With Three-Dimensional Learning

Although there is inherent value in increasing students' engagement with scientific content, this particular model is not ideal for three-dimensional instruction for several reasons. First, and most important, the underlying premise of an interactive notebook is that the teacher delivers information while the students are responsible for responding to and ultimately learning it. This is opposed to the vision of science teaching and learning as a continual process of obtaining and analyzing data to create and refine models for scientific phenomena—a vision in which students are explorers, not simply recipients. Copying down information, whether as traditional notes or on a graphic organizer, reflects outdated thinking about effective science instruction.

Interactive notebooks also often prioritize scientific content over practices and process, a dichotomy that is not supported by educational research in science. Learning scientific concepts and acquiring new vocabulary to describe these concepts are important components of science education, but they should not be done in isolation. Students need to regularly engage in science practices such as generating questions, designing methods to answer them, analyzing data, and constructing explanations based on evidence. Their notebooks should be a place to do this work, not just to write down content to memorize.

Meeting the Needs of Elementary Students

As a model specifically created for the classroom, interactive notebooks do provide some benefits for students compared with other notebook models. First, adding foldables and other creative elements can capture student interest in a way that other types of notebooks do not. These creative elements help engage students with disciplinary core ideas and scientific content, not just practices. And unlike lab notebooks and field journals, interactive notebooks necessarily include elementary-specific tasks, such as opportunities for vocabulary development and note-taking.

Proponents of interactive notebooks point to the structure and predictability of entries as a benefit for students. To be clear, there is value in structure and some predictability when it comes to teaching students how to organize their thinking and writing. However, the overly structured approach of an interactive notebook can be limiting for you and your students alike. What happens when content from a lesson doesn't fit neatly onto a two-page spread? Can students have the opportunity to personalize their own learning and work in meaningful formats? Where (and when) do students have the opportunity to revisit and revise their thinking? Decades of research have found that learning is a complex, transactional, and often messy process. Any approach to science notebooking should reflect this understanding.

Finally, the interactive notebook model can be a slippery slope toward valuing and rewarding compliance over deep thinking and learning. The input-output model easily leads to a checklist approach to assessment: Are all output assignments complete? Have directions been followed? Are entries neat and organized? Focusing on these types of criteria for assessments distracts from the true purpose of a notebook: to support students' sensemaking of phenomena and scientific content.

I am not implying that there is never a time or place for note-taking, and a well-timed graphic organizer or foldable can certainly boost student understanding. Such elements are reflected in my approach to science notebooks later in this chapter and throughout the entire book. However, I use these tools sparingly, and only when other approaches do not suit the content or my purpose.

Digital Notebook

Although a digital notebook is not a model in and of itself, many teachers (and their students) use software, tools, and apps to create digital science notebooks. It is easy to understand the appeal: Such notebooks are characterized by easier organization, the ability to integrate photographs and data from probes and other electronic measuring tools, and the possibility of universal access thanks to the cloud. Saving paper and appealing to students' interest in all things technology-related are also quite compelling. Moreover, digital notebooks provide access for students with disabilities through audio recording, text-to-speech programs, and other features. Recent research (Paek and Fulton 2016; Fulton, Paek, and Taoka 2017) has demonstrated that elementary students are capable of using a digital notebook successfully, although there are specific developmental needs to consider, such as writing on a tablet versus typing with a keyboard (Paek and Fulton 2014). Additionally, a study using a web-based science notebook based on Universal Design for Learning (UDL) principles found improved science learning outcomes, as well as high levels of student and teacher interest, feelings of competence, and autonomy (Rappolt-Schlichtmann et al. 2013).

While some teachers find digital notebooks to be superior to traditional paper notebooks (Constantine and Jung 2019), they are not a one-size-fits-all solution. Research findings suggest a real benefit to handwriting in physical notebooks, including greater idea generation (Berninger et al. 2006, 2009), better learning of content (Mueller and Oppenheimer 2014), and better memory (Smoker, Murphy, and Rockwell 2009). For some teachers, these benefits outweigh the convenience of digital notebooks. Others do not have the devices needed to support the effective use of a digital notebook platform. And still others might use both in tandem to take advantage of the strengths of each format.

In my mind, the format (paper or digital) matters less than the pedagogical approach being used. The chapters that follow include many examples from paper notebooks, as this is what I have used—and continue to use—in my elementary science classroom. Though I occasionally use technology to create notebook entries and do offer typing as an alternative for students with writing-specific learning differences, most of our notebook work is done by hand. However, the approach to notebooking that I present in subsequent chapters can be applied to both paper and digital formats.

My Flexible Approach to Notebooking

I do not follow any one specific model for my students' notebooks. Instead, I use a more flexible approach in which I borrow and adapt specific strategies and templates from the various models, matching *format* and *task* to the *purpose* of a specific lesson or instructional activity. If my students are designing and conducting an experiment that follows a specific procedure or process, their notebook entries will reflect those of a lab notebook. If we are outdoors in the schoolyard or on a field trip, their entries might resemble those found in a field journal. On the rare occasion that I provide students with content (only if they cannot discover it themselves), I might ask students to respond to this information in a manner similar to an interactive notebook entry. I may choose to include a graphic organizer or foldable if I feel that it will truly enhance student understanding of a concept. I also provide scaffolds and exemplars to help students approach thinking, writing, and speaking in scientific ways. (For more information on how I plan a unit of study with the notebook in mind, see Chapter 9.)

I also strive to balance the number of teacher-directed entries with opportunities for students to use *voice* and *choice* in their science notebooks. To help my students learn to effectively represent and organize information, I choose to include organizational elements such as a table of contents and glossary pages. At times, I model how to set up a particular notebook page, dividing the space into sections or creating a table. This is especially true at the beginning of a school year, with new students, or when my students are working on a new type of entry. However, I make a pointed effort to also allow times when students are free to organize and represent their work in any way

Chapter 3

that makes sense to them. Although this means that entries can be messy and harder to read and assess, I know that the only way students will truly develop these abilities is through continued trial and error. (See Chapter 5 for approaches that help students learn from their peers and improve their notebooking skills, including the use of a document camera and notebook workshop time.)

Finally, it is worth stating that my students' notebooks look different from year to year and even from class to class. While there are elements that I have fine-tuned and use consistently each year, I know that my use of notebooks will be most effective if it is responsive to my students and their individual needs—as well as to my own learning and growth as a teacher. In the chapters to come, I share my most effective techniques to date and examples, knowing that you will adapt my ideas to suit your own particular situation and students.

Various models of science notebooks exist, yet they are not all created equal in terms of their appropriateness for three-dimensional learning and the developmental needs of elementary students. Rather than adopt a singular model, I borrow elements from each as appropriate for the content and task at hand. This flexible approach allows me to tailor my notebooking strategies to individual classes. In the next chapter, I share strategies for effectively introducing science notebooks to your students.

References

Berninger, V. W., R. D. Abbott, J. Jones, B. J. Wolf, L. Gould, M. Anderson-Youngstrom, S. Shimada, and K. Apel. 2006. Early development of language by hand: Composing, reading, listening, and speaking connections; three letter-writing modes; and fast mapping in spelling. *Developmental Neuropsychology* 29 (1): 61-92, *https://doi.org/10.1207/s15326942dn2901_5*.

Berninger, V. W., T. L. Richards, P. S. Stock, R. D. Abbott, P. A. Trivedi, L. E. Altemeier, and J. R. Hayes. 2009. fMRI activation related to nature of ideas generated and differences between good and poor writers during idea generation. *BJEP Monograph Series II, Number 6: Teaching and Learning Writing* 17: 77–93.

Chesbro, R. 2006. Using interactive notebooks for inquiry science. *Science Scope* 29 (7): 30–34.

Collins, A., J. S. Brown, and A. Holum. 1991. Cognitive apprenticeship: Making thinking visible. *American Educator* 15 (3): 6–11.

Collins, A., J. S. Brown, and S. E. Newman. 1989. Cognitive apprenticeship teaching and the crafts of reading, writing, and mathematics. In *Knowing, learning, and instruction*, ed. L. B. Resnick, 453–494. Hillsdale, NJ: Lawrence Erlbaum Associates.

Constantine, A., and K. G. Jung. 2019. Using digital science notebooks to support elementary student learning: Lessons and perspectives from a fifth-grade science classroom. *Contemporary Issues in Technology and Teacher Education* 19 (3): 373–412. Waynesville, NC: Society for Information Technology and Teacher Education. *www.learntechlib.org/primary/p/207107/*.

Fulton, L. A., S. Paek, and M. Taoka. 2017. Science notebooks for the 21st century. *Science and Children* 54 (5): 54–59.

Jason, A. 2012. *Mythbusters*. Episode 186, "Bouncing bullet." Aired May 13 on Discovery Channel. *https://mythresults.com/bouncing-bullet*.

Mueller, P. A., and D. M. Oppenheimer. 2014. The pen is mightier than the keyboard: Advantages of longhand over laptop note taking. *Psychological Science* 25 (6): 1159–1168.

National Governors Association Center for Best Practices and Council of Chief State School Officers (NGAC and CCSSO). 2010. *Common core state standards*. Washington, DC: NGAC and CCSSO.

National Research Council (NRC). 2007. *Taking Science to School: Learning and Teaching Science in Grades K–8*. Washington, DC: National Academies Press. *https://doi.org/10.17226/11625*.

National Research Council (NRC). 2010. *Surrounded by science: Learning science in informal environments*. Washington, DC: National Academies Press. *https://doi.org/10.17226/12614*.

Paek, S., and L. A. Fulton. 2014. Digital science notebooks to support elementary students' scientific practices. Paper presented at the Association for Educational Communications and Technology International Convention, Jacksonville, FL, November. *https://members.aect.org/pdf/Proceedings/proceedings14/2014i/14_21.pdf*.

Paek, S., and L. A. Fulton. 2016. Writing in digital science notebooks: What features do elementary students use? *International Journal* 10 (2): 44–51.

Rappolt-Schlichtmann, G., S. G. Daley, S. Lim, S. Lapinski, K. H. Robinson, and M. Johnson. 2013. Universal Design for Learning and elementary school science: Exploring the efficacy, use, and perceptions of a web-based science notebook. *Journal of Educational Psychology* 105 (4): 1210.

Robinson, C. 2018. Interactive and digital notebooks in the science classroom. *Science Scope* 41 (7): 20–23, 25.

Smoker, T. J., C. E. Murphy, and A. K. Rockwell. 2009. Comparing memory for handwriting versus typing. *Proceedings of the Human Factors and Ergonomics Society Annual Meeting* 53 (22): 1744–1747.

Young, J. 2003. Science interactive notebooks in the classroom. *Science Scope* 26 (4): 44–47.

CHAPTER 4

Launching Science Notebooks With Students

I am proud of having a notebook like a real scientist.

—First-grade student, on end-of-year survey about science class

Science notebooks are dynamic tools that serve as a place for students to make sense of ideas and take ownership of their learning. Highly personalized, the science notebook is a valuable record of the intellectual growth a student has made over the course of a year. But how do you keep this notebook special amid a sea of classwork, homework, and other curricular responsibilities? How do you help students feel that this notebook is more than just another assignment to complete? In this chapter, I share activities and strategies I use to help my students take pride in their notebooks.

As an elementary science specialist, I have had the pleasure and privilege of working with the same students over a five-year period. Although we welcome new students to all grades each year, the way I set the stage in first grade has implications for how my students view science in fifth grade. With this in mind, I carefully craft how my first-grade students are introduced to the habits of mind and tools of a scientist, including their science notebooks. I hope that this introduction proves useful for your elementary students, even if they aren't in first grade.

The Most Important Scientist

My opening unit with first-grade students is an introduction to the tools, skills, and types of activities they will engage in over the next five years in my class. Students practice using hand lenses, microscopes, and balances within the context of simple investigations; they make predictions and record observations and simple forms of data. As a class, students learn how to participate in the "scientist meetings" (class discussions) that I hold after each investigation and how to be safe during all types of activities. But there's much more to my unit than that.

I am deeply concerned with building students' identities as scientists, and my introductory first-grade unit is my chance to set the stage for the work I will continue with them throughout elementary school. I want all my students to view themselves as scientists, part of a global and historic community that engages in specific types of investigation to study and understand the natural world. To this end, I introduce my students to scientists and explicitly talk about the actions and habits of mind of scientists on a regular basis.

Chapter 4

In this first-grade unit, students listen to read-alouds of picture book biographies about scientists such as John James Audubon and Jane Goodall as they practice observing and drawing. Table 4.1 includes a sampling of my favorite picture book biographies to share with students, but feel free to modify this list to meet your own instructional needs. They learn about other scientists through a variety of portraits of women in science that hang around my classroom (Figure 4.1). In addition to briefly sharing about each scientist's life and work, I make connections to the skills that students are developing in my classroom activities.

Table 4.1. Suggested picture book biographies to introduce students to scientists and people in STEM.

Scientist	Title and author	Highlighted practices
Patricia Bath	*The Doctor With an Eye for Eyes: The Story of Dr. Patricia Bath* by Julia Finley Mosca	Asking questions and defining problems
Marie Tharp	*Solving the Puzzle Under the Sea: Marie Tharp Maps the Ocean Floor* by Robert Burleigh	Developing and using models
Marie Tharp	*Ocean Speaks: How Marie Tharp Revealed the Ocean's Biggest Secret* by Jess Keating	Developing and using models
Charles Henry Turner	*Buzzing With Questions: The Inquisitive Mind of Charles Henry Turner* by Janice N. Harrington	Planning and carrying out investigations
John James Audubon	*The Boy Who Drew Birds: A Story of John James Audubon* by Jacqueline Davies	Planning and carrying out investigations (specifically, observations)
Jane Goodall	*The Watcher: Jane Goodall's Life With the Chimps* by Jeanette Winter	Planning and carrying out investigations (specifically, observations)
Wu Chien Shiung	*Queen of Physics: How Wu Chien Shiung Helped Unlock the Secrets of the Atom* by Teresa Robeson	Analyzing and interpreting data
Katherine Johnson	*Counting on Katherine: How Katherine Johnson Saved Apollo 13* by Helaine Becker	Using mathematics and computational thinking
Margaret Hamilton	*Margaret and the Moon: How Margaret Hamilton Saved the First Lunar Landing* by Dean Robbins	Using mathematics and computational thinking
Temple Grandin	*The Girl Who Thought in Pictures: The Story of Dr. Temple Grandin* by Julia Finley Mosca	Constructing explanations and designing solutions
Eugenie Clark	*Shark Lady: The True Story of How Eugenie Clark Became the Ocean's Most Fearless Scientist* by Jess Keating	Engaging in argument from evidence
Rachel Carson	*Rachel Carson and Her Book That Changed the World* by Laurie Lawlor	Obtaining, evaluating, and communicating information

Note: While I have aligned each book to a particular science and engineering practice, it is important to note that most books will target many SEPs. For a more complete list, use QR Code 4.1.

QR Code 4.1. Blog post with list of picture book biographies aligned to the science and engineering practices.

Figure 4.1. Student-painted portraits of notable women in science hang in my classroom.

Near the end of the unit, I ask students to share the names of some of the scientists they have learned about. After a brief discussion, I tell students that I have another scientist for them to meet—and that this person might be the most important scientist of all. I produce a colorfully wrapped shoebox and explain that this person's picture is inside the box. Students will each take a turn to peek inside and see the scientist's picture. I reassure students that they will recognize this person and ask them not to blurt out the name of the scientist while others are waiting their turn. Then I present the box to each student in turn, holding the lid open so that the student may peer inside. Giggles usually erupt almost immediately.

What's inside the box? A mirror!

Once everyone has had a turn, I ask students to share whom they saw inside the box and am always met with a resounding chorus of "Me!" We talk about how they indeed are all scientists, drawing from the experiences that they have had in class thus far and any other knowledge they have of science and scientists. Students finish this lesson by drawing a self-portrait and completing the sentence starter "I am a scientist because …" (see Figure 4.2 on p. 38 for one example). These are displayed on a bulletin board to welcome the newest members of our scientific community.

Figure 4.2. A first-grade student's self-portrait and explanation of why the student is a scientist.

Figure 4.3. Reading aloud *Notable Notebooks: Scientists and Their Writings* is a great way to tie students' notebooks to the work of actual scientists.

In recent years, I've started the year with older grades getting another peek inside the shoebox. I frame this as letting our new students meet the most important scientist, and I've found that my returning students not only giggle conspiratorially but also want to look inside again for themselves. In my experience, even fourth and fifth graders secretly delight in seeing their own reflection inside that shoebox, even if they try to act otherwise. Starting each year with a reinforcement of this scientific identity is a small but important piece of helping my students truly view themselves as scientists.

This is not the only time I refer to my students as scientists; in fact, it is a term I use to address them regularly throughout their time in my class. I hope that by hearing this description of themselves routinely, students internalize this as a valued piece of their identity. This instructional sequence sets the stage for the introduction of science notebooks.

Introducing the Science Notebooks

I introduce science notebooks as a tool by reading and discussing my picture book *Notable Notebooks: Scientists and Their Writings* (Figure 4.3; Fries-Gaither 2016). In fact, this is exactly the situation that inspired me to write the book in the first place! The book highlights nine historical and contemporary scientists—Galileo, Sir Isaac Newton, Beatrix Potter, Maria Mitchell, Charles Henry Turner, Lonnie Thompson, Jane Goodall, Ellen Ochoa, and Stephanie Kwolek—and the central role notebooks have played in each of their work. With first-grade students at the start of the year, I often choose to read the opening pages, a select few scientists (such as Jane Goodall), and the closing page. If I am introducing science notebooks to older students, I read the book in its entirety. I've also found that students love watching the video of *Notable Notebooks* being read aloud by astronaut Joseph Acaba aboard the International Space Station (see QR Code 4.2 for the link).

QR Code 4.2. *Notable Notebooks* read aboard the International Space Station through the Story Time From Space program.

After reading and discussing *Notable Notebooks*, I explain to students that because they, too, are scientists, they need to have their own notebooks. I pass out notebooks to students and have them write their names on the covers. Some teachers give students time to decorate their notebook covers with science-themed pictures and stickers. Although I have never done this because of time constraints, it may help students take ownership of this tool.

Once the notebooks have been distributed, it's important to have your students start using them regularly. Depending on your instructional needs, you may have students add organizational elements to the notebook before launching into content-based investigations. I have primary students glue in alphabet box graphic organizers to keep track of vocabulary, while older students add a table of contents and number the notebook pages. (For more organizational information, see Chapter 5.)

Don't despair if some of your students are not immediately enamored with the idea of keeping a notebook. Even after your careful introduction of the tool, some students, particularly reluctant writers, may be less than enthused at first. Keep the tone of assignments invitational, focus on strengths in your students' entries, and provide opportunities for student reflection. Witnessing growth through notebook entries is a source of pride and a powerful motivator for students. These types of instructional strategies are discussed in more detail throughout the subsequent chapters of this book.

Students of all ages benefit from thoughtful, explicit introductions to scientific tools and skills, and you can adjust the context around this work to match their developmental levels and whatever content begins your year.

Creating Your Own Lessons to Introduce Notebooks

If you'd like to craft your own series of lessons or unit to introduce notebooks to your students, keep the following general principles in mind.

Introduce Students to Historical and Contemporary Scientists

It is important for students to learn about the incredible diversity of people who have contributed to the scientific enterprise over time. We are fortunate to live in a golden age of picture book biographies, which make the stories of many scientists accessible to children. Find ways to weave these texts into your teaching, whether it is in conjunction with a particular disciplinary core idea, in a study focused on the nature of science, or even in stand-alone read-alouds. Self-contained classroom teachers might choose to focus on scientists in a biography study, incorporating science, English language arts, and even history content into a single unit. Others might choose to

Figure 4.4. *Finding Wonders: Three Girls Who Changed Science* is a wonderful read-aloud for upper elementary or middle school students.

Figure 4.5. *Inquiring Scientists, Inquiring Readers* includes a unit plan about scientists and their work.

use read-aloud time for this work. *Finding Wonders: Three Girls Who Changed Science* by Jeannine Atkins (2016), a novel in verse that profiles Maria Merian, Mary Anning, and Maria (ma-RYE-ah) Mitchell, is a terrific read-aloud or independent read for upper elementary (and even middle school) students that not only introduces students to the work of these young women but also describes the historical context that made their discoveries so remarkable (Figure 4.4).

In addition to the ideas described in this chapter, the unit "Scientists Like Me" in *Inquiring Scientists, Inquiring Readers: Using Nonfiction to Promote Science Literacy, Grades 3–5* (Figure 4.5; Fries-Gaither and Shiverdecker 2012) provides a learning-cycle framework for introducing the work of scientists in conjunction with scientific practices such as observing, planning and conducting experiments, and writing. This framework could be revised or expanded to include other scientists, as many new excellent picture book biographies are now available.

Explicitly Teach Scientific Practices and Habits of Mind

Science educators have long engaged their students in lessons and activities that require them to ask questions, plan and conduct investigations, work with data, and communicate findings. Once referred to as process skills (Padilla 1990), these have been revised and refined into eight science and engineering practices (NRC 2012). (These are discussed in greater detail in Chapters 1 and 6; see Figure 1.1 on p. 3 for a list of the SEPs.)

Over the years, I have learned that these practices don't come naturally to students—they must be taught. Students need explicit instruction in the *whens* and *hows* of these practices, and they need time and repeated attempts to gain proficiency in them. Additionally, students often benefit from hearing the mental thought processes that underlie the practices. Just as a reading teacher might use a think-aloud to help students understand what is happening in the teacher's mind as the teacher reads and applies comprehension strategies, so might a science teacher do likewise as the teacher is analyzing data or identifying the strongest evidence for an argument.

What does this have to do with notebooking? Keeping a science notebook naturally means engaging in these types of work. As students begin a notebook, this is an opportune time to introduce one or more of the practices to them, with the goal of revisiting each repeatedly throughout the school year. The more intentional and deliberate you are with these lessons, the more successful students will be. And the more successful they are, the more engaged they will be with their notebooks. It's a positive feedback loop. (For examples of instructional routines and notebook entries that support the development of the science and engineering practices, see Chapter 6.)

Connect Students' Work to the Work of Scientists

The final guideline is to take every opportunity to make connections between the work students are doing in the science classroom and the work of the historical and contemporary scientists they've studied. As I plan my units of study, I try to include at least one scientist whose work aligns in some way with what my students are doing. For example, when students are engaged in analyzing the locations of landforms and other features around the world to identify patterns, I read *Solving the Puzzle Under the Sea: Marie Tharp Maps the Ocean Floor* by Robert Burleigh (2016) and *Ocean Speaks: How Marie Tharp Revealed the Ocean's Biggest Secret* by Jess Keating (2020) (Figure 4.6) and help students reflect on how their own work closely mirrors that of Tharp and her colleague Bruce Heezen.

Figure 4.6. Examples of read-alouds that connect students' work in the classroom to the work of historical and contemporary scientists.

(a)   (b)

Solving the Puzzle Under the Sea: Marie Tharp Maps the Ocean Floor (a) and *Ocean Speaks: How Marie Tharp Revealed the Ocean's Biggest Secret* (b).

Keeping a notebook with observations, data, and ideas isn't just an assignment dreamed up by science educators—it is at the heart of professional scientists' work. Students feel empowered and are more likely to engage when they see the work they are doing as productive and meaningful. In this chapter, I have shared how I introduce my students to science notebooks and provided guidelines to help you plan your own lessons to do the same. In the next chapter, I introduce elements and processes that help my students learn to keep their notebooks organized.

References

Atkins, J. 2016. *Finding wonders: Three girls that changed science.* New York: Atheneum Books for Young Readers.

Burleigh, R. 2016. *Solving the puzzle under the sea: Marie Tharp maps the ocean floor.* New York: Simon & Schuster.

Fries-Gaither, J. 2016. *Notable notebooks: Scientists and their writings.* Arlington, VA: NSTA Press.

Fries-Gaither, J., and T. Shiverdecker. 2012. *Inquiring scientists, inquiring readers: Using nonfiction to promote science literacy, grades 3–5.* Arlington, VA: NSTA Press.

Keating, J. 2020. *Ocean speaks: How Marie Tharp revealed the ocean's biggest secret.* Toronto: Tundra Books.

National Research Council (NRC). 2012. *A framework for K–12 science education: Practices, crosscutting concepts, and core ideas.* Washington, DC: National Academies Press. *https://doi.org/10.17226/13165.*

Padilla, M. J. 1990. The science process skills. *Research matters—To the science teacher,* No. 9004. Reston, VA: NARST. *https://narst.org/research-matters/science-process-skills.*

CHAPTER 5

Notebook Organization

Science is organized knowledge.

—Herbert Spencer, "The Art of Education"

Our goal as science educators is to cognitively apprentice students into the scientific discipline (Collins, Brown, and Newman 1989; Collins, Brown, and Holum 1991), while at the same time taking into account the important differences between school science and science as a discipline and the developmental needs of our students. So in addition to providing students with space and opportunity to make sense of scientific phenomena, science notebooks allow for teaching and modeling of organizational skills and the use of nonfiction text features. This is the context and rationale for the way I approach science notebooks in my classroom, blending the pure scientific content with elements and strategies that cross disciplines and help build executive functioning skills. In this chapter, I explain the elements of my science notebooking approach that build organization skills, as well as a gradual release of responsibility model that I use to balance scaffolding student work with affording opportunities for independence and choice. This chapter is not meant to provide a prescriptive list; rather it provides some suggestions for how you might organize notebooks in your own classroom. Just as I have tried and revised various features over the years (and continue to do so based on the needs of my current students), you should select an approach that works for your specific context.

The Basics

I prefer the classic black-and-white composition notebooks over spiral-bound books; I've found the metal spiral can be easily unwound and can catch on other notebooks, making it difficult to separate and pass out notebooks quickly at the beginning of the class. For my youngest students (K–2), I use a primary-ruled version of the composition book both to be consistent with the paper they use in class and to accommodate their need for larger spaces in which to write.

One downside to the composition books is that they are virtually indistinguishable from one another, which can slow down their distribution at the beginning of class. My colleague and fellow science teacher had the brilliant idea to label the spines and tops of all notebooks with student names (Figure 5.1, p. 44). This allows students to spot their notebooks more easily and minimizes the time needed to get supplies before class can start.

Figure 5.1. Having student names on notebook tops and spines makes it easier for students to find their notebooks at the beginning of class.

(a)

(b)

Student names written on notebook tops (a) and spine (b).

Students store their notebooks in my classroom rather than carry them back and forth between their homeroom and the science lab. Storing over 150 notebooks requires a dedicated and accessible space. I dedicated a portion of one counter to notebooks and have a box for each homeroom class. Students pick up their notebooks at the beginning of class and put them away as they line up to leave at the end of the period.

General Notebook Organization

My students begin each new year with a fresh notebook, even if they hadn't completely filled the notebook from the previous year. As this notebook serves as a cumulative record of what students have learned, I prefer to have each school year contained in its own volume. Additionally, it is helpful to have all my students quite literally start on the same page at the beginning of the year, particularly for new students.

Figure 5.2. A ribbon hot glued into the notebook serves as a bookmark and helps students keep their place.

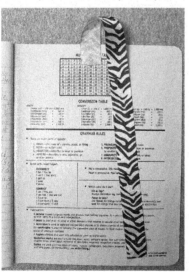

I add one organizational tool to notebooks before students ever use them: a fabric ribbon bookmark hot glued onto the inside back cover (Figure 5.2). I shop the clearance racks at craft stores all year, stockpiling discounted ribbon in a variety of colors and patterns. I've found that for a standard-size composition book, 12 to 13 inches of ribbon is a suitable length. Coating the free end of the ribbon with a small amount of clear nail polish prevents fraying, although I always keep extra ribbon on hand for students who still manage to pick at and fray their bookmarks.

When students receive their notebooks, I have them each write their name and class on the front of the notebook and start adding further organizational tools. For my students in third grade and up, that means numbering their notebook pages and adding a table of contents. (I haven't had much success with having first or second graders number notebook pages or use a table of contents. For the younger grades, I simply remind them to use their bookmarks to mark their place and use dated entries to maintain a sense of organization.) Students number the front and back of each page, as shown in Figure 5.3. I model where students may place the numbers and give them a choice of the top or bottom corners of the pages as long as they are consistent. I've found that this task works best when divided into small chunks of time spread out over a period of several days or a week; the first few and last few minutes of class are perfect for this. This way, the thought of numbering 200 pages doesn't seem so daunting. Students use their bookmarks to keep track of their place while numbering and work at their own pace. I support students who need help and encouragement, and I troubleshoot as needed if students accidentally skip some pages. My third graders have needed a fair amount of support during notebook setup at the beginning of the year, but the time spent pays off the following year, when the organizational system is more familiar.

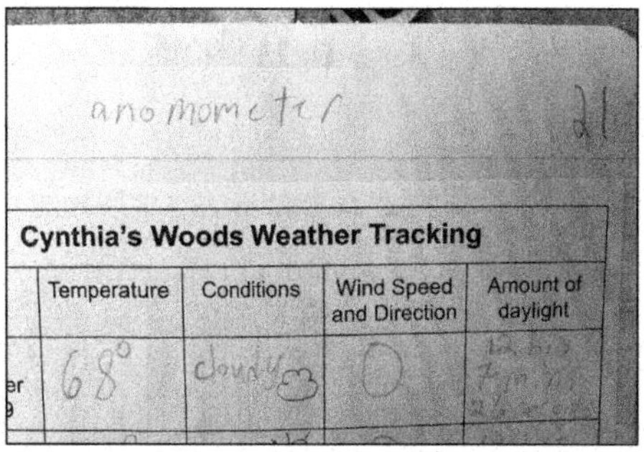

I have students number their pages because it allows them to use a table of contents in the notebook—an element that proves invaluable throughout the year. Throughout the elementary grades, students learn about nonfiction text features such as a table of contents, an index, and a glossary, and I've found that creating some of these features deepens student understanding of what they are and how they are used. After many iterations and revisions, I have settled on a preprinted table of contents that is folded in half width-wise and glued onto the first page of students' notebooks (see Figure 5.4). This eliminates the guesswork of how many pages to reserve for an element that students add to throughout the year, and it provides scaffolding for students as they gain confidence in creating a table of contents. (A copy of this table of contents is provided for your convenience in the appendix on pp. 112–113.)

Safety Contract

Safety is an important topic for science educators, and I want my elementary students to take this just as seriously as older students do. To that end, one of my very first classes focuses on ways to safely engage with materials and with others. I vary the actual lesson from year to year: Sometimes we have a class conversation to generate a list of safety rules, and other times students brainstorm rules and vote for their top choices as a class. Still other times, I task each table group with the creation of a set of safety rules. No matter the format, the ultimate goal and product are the same: a mutually agreed-upon safety contract for the year (Figure 5.5, p. 46).

Figure 5.3. Page numbering in notebooks keeps intermediate students organized.

Figure 5.4. A table of contents makes it easier for students to find items in their notebooks.

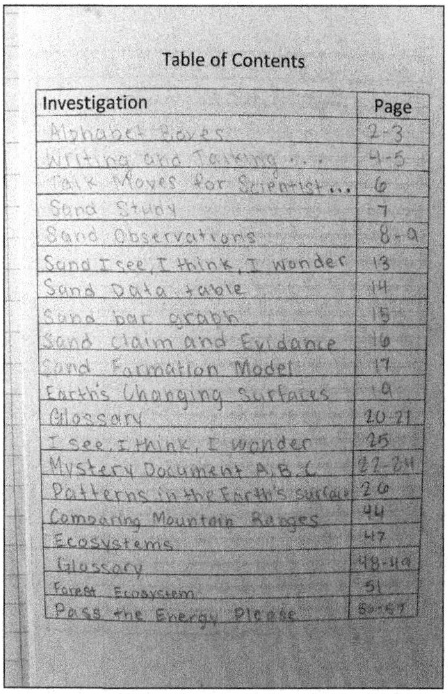

Chapter 5

Figure 5.5. Various safety contracts created with students over the years.

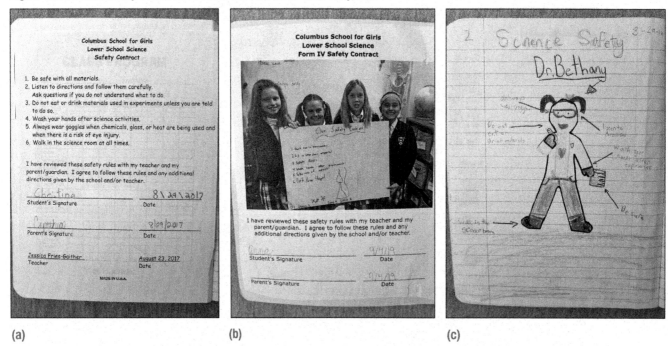

(a)   (b)   (c)

A whole-class contract (a); small-group contracts (b); and a drawing created to accompany the safety contract (c).

I make copies of the safety contract and send them home with students to read and discuss with their parents and sign. When students bring their signed contracts back, we cut them down to size and glue them into the inside front cover of their notebooks (Figures 5.5a and 5.5b). Students see their contracts every time they open their notebooks, keeping safety at the forefront of their minds. Some years, I have also asked students to draw a picture of themselves following the safety rules (Figure 5.5c) to promote metacognition and greater ownership of the safety contract. (Sample safety contracts for primary and intermediate students are included in the appendix on pp. 114–115 for your use.)

Vocabulary Elements

Although thought about best practices with regard to vocabulary instruction and development has changed over time (see Chapter 2 for a greater discussion of this topic), I still include vocabulary-specific elements in my notebooks. The first element is alphabet boxes, which are glued to the first and second pages of my primary students' notebooks or immediately following the table of contents in my intermediate students' notebooks. Students record new vocabulary words in this organizer throughout the year, providing an at-a-glance snapshot of the content studied as well as easy-to-access support for spelling these new words (Figure 5.6). While the majority of words are added as a class and match the terms on our word wall, I also encourage students to add personal words to their alphabet boxes.

Figure 5.6. Alphabet boxes serve as an organizer in which students record new vocabulary words.

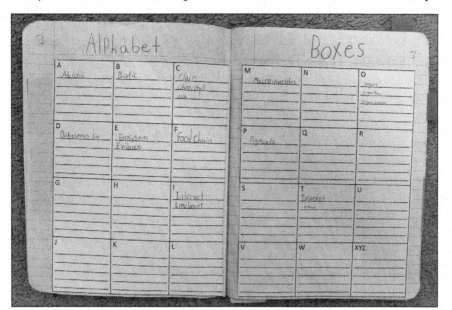

The second vocabulary element, a glossary, is used only by my intermediate students in third grade and up. As with the table of contents, creating this text feature helps students better understand the purpose of glossaries in published nonfiction, and it has also gone through many iterations in my own notebooking practice. Instead of a large glossary at the very end of the notebook, which proved to be difficult in terms of knowing how many pages to reserve, I have my students add smaller, unit-specific glossaries throughout the notebook.

To provide consistency and scaffolding in this important section, I use photocopied teacher-created pages for the glossary pages. Each glossary entry includes the word, a student-generated definition, and a simple picture or symbol that represents the word (Figure 5.7). I try to anticipate how many pages to use for core vocabulary terms, but I also include a few extra blank rows for words that emerge from class discussions and student questions.

While the structure of the glossary is straightforward, some implementation points are worth discussing further. Science vocabulary represents concepts, and students must understand those concepts before being expected to master

Figure 5.7. The glossary is created with student-generated definitions and pictures.

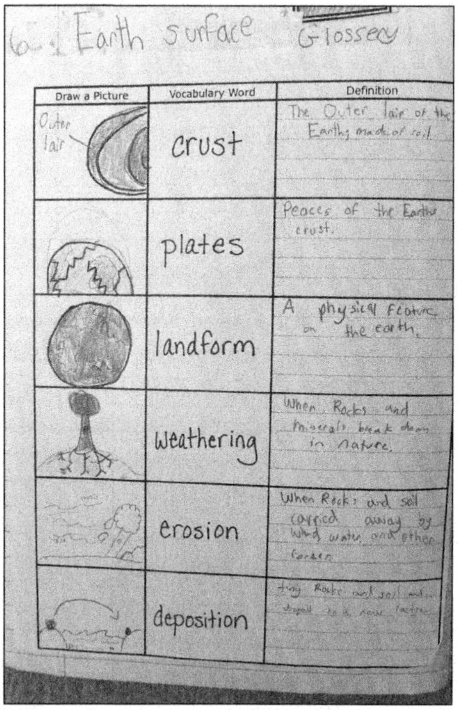

the use of a given term. So although I have students add the glossary pages to their notebook early on in a unit, they typically do not add words, definitions, or pictures until much later in the study. Only once I am confident that students have developed some conceptual understanding through investigation and activity do I ask them to work on their glossaries. When the time comes, I tend to challenge students to generate their own definitions, either individually or in small groups. We discuss student ideas and typically settle on a definition that incorporates thinking from many individuals or groups. Students then each add their own picture or symbol to represent the term. These pictures provide interesting insight into students' understanding of the concepts behind the vocabulary. For example, a fifth grader's picture of a graveyard to represent the word *abiotic* (Figure 5.8) told me that the student was still working through the difference between something that is dead and something that was never alive in the first place.

Figure 5.8. A fifth-grade student's sketch for the word *abiotic* reveals a misunderstanding of the term.

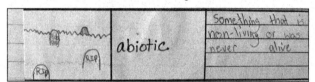

Another note concerns the number of words to include in a given unit's glossary. Science is a terminology-heavy discipline, and vocabulary lists for units or chapters used to be quite extensive, but that approach does not lead to deep understanding. Instead, I choose a limited number of core words that encapsulate the foundational concepts of the unit to include in the glossary, using language from standards documents and other sources as a guide. Other words may be introduced and used as needed, but they will not be emphasized in the same way as the words that make up the unit glossary. Of course, additional words emerge through student-directed conversations, but after a number of years teaching the same topics, it becomes easier to predict which words will be central to a unit.

Vocabulary development and support are embedded into other notebook entries besides alphabet boxes and the glossary. Those two features simply make up the formal structure for vocabulary in the notebook. (For discussion of other types of notebook entries that build vocabulary, see Chapter 6.)

Unit Organization

I divide my curriculum into several discrete units of study, and this organization is mirrored in my older students' notebooks (third grade and up). Each unit is organized in the same way, although the contents of each unit vary widely depending on the investigations and instructional activities used. Students in first and second grades work sequentially through their notebooks without additional organizational elements, as I have found these to be more of a distraction than a help in the early grades.

The unit begins with a cover page that includes the title of the unit, student illustrations or brainstorming, and a tab (Figures 5.9 and 5.10). This cover page is added to the table of contents as the first page of the unit of study. The title of the unit can be a single word (Ecosystems), a phrase (Human Body Systems), or even a guiding question (What's Happening to Our Corn?). I ask students to decorate the page with colorful illustrations related to the title of the unit. They enjoy this chance to personalize their notebooks, and their drawings give me an idea of the understandings they bring with them to the study. For example, one student's illustrations tell me that she is quite familiar with examples of electrical energy but may not know that there are other forms of energy as well (Figure 5.9). The tab is an easy way for students to jump to the correct unit, especially late in the year when a good chunk of the notebook has been filled. I help students create tabs by folding 2-by-2-inch-square sticky notes in half and attaching the two free ends to the front and back of the cover page (Figure 5.10). We add some glue to the inside of the folded sticky note to give it staying power. Bright colors are eye-catching and differentiate one unit from the next.

The unit glossary follows the cover page, and this is also added to the table of contents. After the glossary, notebook entries vary depending on the instructional sequence of the unit. (Many of these strategies and entries are discussed in detail in Chapter 6.)

Individual Entry Organization

The final level of organization that I ask students to include in their notebooks is a system for their entries (Figure 5.11). Each time we start a new investigation or instructional activity, I direct students to include a title and the date at the top of the page. This title is used in the table of contents and reinforces the concept of section headings in nonfiction text. Dating entries also mirrors the practice of scientists in their notebooks. Students need guidance at first, but most quickly adopt this as a regular habit.

Doodle Pages

The final section of my students' notebooks is not instructional at all. Instead, it emerged from a challenge I faced early in my implementation of notebooks—students doodling all over their notebook pages, which made their work difficult to read. Despite my frustration and repeated reminders, students kept right on doodling. After some reflection on my own tendency to doodle during meetings and while on phone calls, I decided to give students a few pages at the back of the notebook as a

Figure 5.9. This fourth-grade student's illustrations on her unit cover page suggest that she knows about electrical energy but may not be familiar with other forms of energy.

Figure 5.10. Unit tabs make it easy to find the current unit of study.

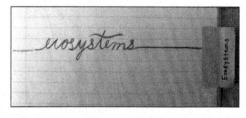

Figure 5.11. Each notebook entry heading includes a title and the date.

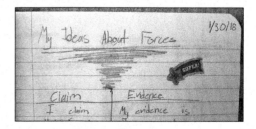

Figure 5.12. Doodle pages at the back of the notebook give students a place to draw while keeping their work pages clean.

place to doodle (Figure 5.12). I usually allow the last three to five pages to be used for doodling, and I've never run into a situation where those pages were missed for academic work. At the beginning of the year, I discuss appropriate times to use these pages (e.g., before the start of class, after finishing an assignment), and I remind individual students of these guidelines as needed. While it hasn't eliminated doodling on the work pages entirely, it has dramatically improved the neatness and readability of notebook entries.

Scaffolding for Student Independence

At this point, it may seem as though I provide students with many more premade organizers and elements than you initially expected. I admit that I do prefer teacher-created *organizational elements* in my notebooks, as I find it levels the playing field for students in terms of organizational skills. Some of my students have a natural proclivity for neat and organized work, and they would undoubtedly have exceptionally organized notebooks no matter what approach I used. But for many others, organizing written work is a skill that needs to be developed. Without these premade and teacher-directed elements, their notebooks would quickly become disorganized and difficult for the students to navigate. They would also be incredibly frustrating to review as a teacher. Therefore, for setting up the *overall structure and format* of the notebook, I prefer to use a more teacher-directed approach. My hope is (and experience has confirmed) that my students tend to internalize these structures over time and become more proficient in independently organizing their work.

In terms of content and investigation-based entries, however, my approach is different. As with many other aspects of teaching, I view the use of a notebook on a continuum from teacher-directed to student-directed. As shown in Figure 5.13, a teacher-directed approach to science notebooks includes prescriptive formats for entries; consistent use of teacher-created charts, tables, and organizers; and

Figure 5.13. Science notebook continuum.

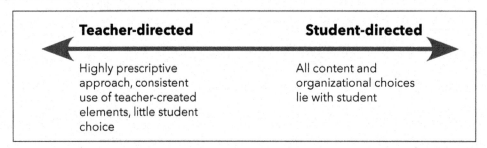

highly directed instructions. On the opposite end of the continuum, a student-directed approach means that all organizational and content decisions lie with the student. In between lies a blend, with some teacher direction when needed, teacher modeling to support students as they grow, and student choice whenever appropriate. My use of science notebooks generally falls into this middle range, as I work to help students investigate science concepts, learn principles and vocabulary, and develop their literacy, organization, and executive functioning skills.

I also think about my students' use of their notebooks with another model in mind—the gradual release of responsibility (GRR) model. This model was originally conceived by literacy experts Pearson and Gallagher (1983). It refers to a structured and intentional pedagogical process in which a teacher first demonstrates a skill or process and then gradually turns over the ownership of that process to the students. The GRR model often includes four steps: modeling or demonstration, shared demonstration, guided practice, and independent practice. It relies on appropriate instructional scaffolding to move students from reliance on the teacher to independence (Figure 5.14).

Figure 5.14. The gradual release of responsibility (GRR) model.

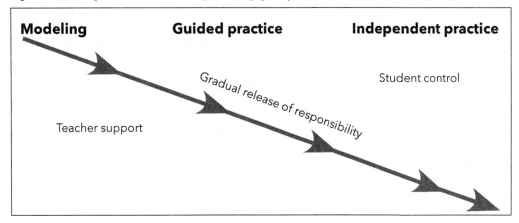

Source: Pearson and Gallagher (1983).

Whereas the original intent of the gradual release of responsibility model was to use it over the course of a lesson or series of lessons, I broaden the concept to include my students' work in their science notebooks over the course of a year or even multiple years. This means that I am more likely to start the school year off with teacher-created charts, graphic organizers, and tables for students to glue into their notebooks. Next, I move to modeling the creation of these elements by drawing them on the board or chart paper and asking students to re-create them in their notebooks. With time and repeated practice, students begin to create their own elements while I provide over-the-shoulder, individual guidance as they work. Eventually, students become proficient and are able to create these types of entries independently. As a science specialist with the luxury of teaching students for multiple years, I also use the GRR model to provide appropriate support for each grade level. My youngest students use more premade components (and for longer in the year) than do my older students, and I am able to build on students' skills from year to year. Whenever possible, though, I let students try their hand at creating and formatting their

Chapter 5

own work. Because a notebook is intended as a personal space for the development of scientific knowledge and practices, imposing too much structure can be limiting. As I plan the day-to-day lessons in my classroom, I constantly attempt to balance the need for scaffolding and guidance with the opportunity for students' voice and choice.

This balance and release of responsibility mean that students don't always get it right the first (or second, or third ...) time. Giving students freedom to create their own notebook entries means that I inevitably see my fair share of messy pages, tables that don't quite make sense, and data-recording methods that seem to be completely unorganized. These entries are not failures, but rather evidence of learning in action. There is great value in giving students the space to try out different organizational techniques, methods of data recording, and elements of observational drawings and modeling; to reflect on what was and wasn't effective; and to revise (or try a new technique in the future). I learn a lot from these entries about how to best support individual students as they gain proficiency with skills and practices (for more discussion on supporting students, see Chapter 7). If you always provide these elements for students because you don't want their work to be messy or because it just seems easier, then you run the risk of having students never actually learn to create these elements in the first place.

So how do I help students improve these types of entries? Individual over-the-shoulder feedback during class time is a highly effective method. But even though I have much to teach students, they also have quite a bit to learn from each other. Knowing this, I have built-in times in my class for students to view the work of their peers, provide feedback, and identify elements that they might like to try in their own work. I use several different techniques to this end, including notebook workshops, gallery walks, and formal and informal partner shares.

Notebook Workshop

Notebook workshop, my own term, is loosely based on the concept of writing workshop popularized by literacy researchers like Lucy Calkins (2006). In writing workshop, the teacher (as mentor author) gives a mini-lesson followed by extended time for students to write independently. During this time, the teacher circulates and coaches students on an as-needed basis. The workshop ends with the sharing of student work.

As a science specialist, I have short class periods and don't have the luxury of the long periods of active work time described in Calkins's writing workshop. So I have created a general framework that I can adapt to a variety of situations and student needs. While I've modified the structure and timing of writing workshop, the core components are still there: mini-lessons, work time with individual feedback and coaching, and sharing of student work. I typically use this model in response to a need I see in a majority of a class's entries. To paint a better picture of what this actually looks like in my classroom, I'll describe two examples of notebook workshop that I use consistently from year to year: observational drawing workshop and claim and evidence workshop. Although these workshops are incredibly helpful, they are not a definitive, one-shot fix; the workshops are often repeated at different points throughout the year.

Observational Drawing Workshop

The first notebook workshop topic that I often use with my students focuses on observational drawing. Early one year, I noticed that my third-grade students' observational sketches of their mealworms (from our investigation of 3-LS1-1: Develop models to describe that organisms have unique and diverse life cycles but all have in common birth, growth, reproduction, and death) were not meeting my expectations. Collectively, their drawings were small, messy, and lacking detail. I decided to spend our next class period in a notebook workshop instead of moving forward with the investigation. I reviewed my students' notebook entries, photographed work that exemplified my criteria for excellent observational drawings, and created a slide show with these examples (see Figure 5.15).

Figure 5.15. Slides with student work samples help students identify essential characteristics of observational drawings.

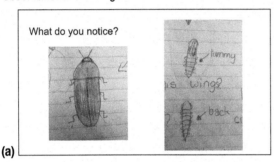
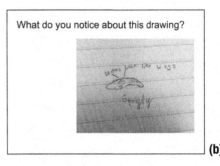

Teacher-created slides showing third-grade students' drawings of mealworms that exemplified criteria for observational drawings, including detailed drawings (a) and the use of labels (b).

In the next class, I set the purpose for the lesson: to improve the quality of students' observational drawings. We discussed why it was important to be able to capture observations in a sketch of an animal or plant and how detailed observations lead to questions that can be investigated. (For more on this connection between observations and questions, see Chapter 6.) I told students that I would be sharing (anonymously) with them a series of pictures of their classmates' work and that their task was to identify why I chose to highlight that particular drawing. As I showed the slides in succession, students discussed the work in table groups before sharing with the larger class. Together, we generated a list of criteria for an excellent observational drawing (Figure 5.16). Fortunately, the students' list matched mine exactly, assuring me that I had chosen the correct work samples to highlight.

Once the list was complete, it was time for students to return to their mealworms. I asked them to create a new observational drawing using the criteria we had generated as a class. I provided

Figure 5.16. Student-generated (with teacher guidance) list of criteria for an excellent observational drawing.

Name: _____		
Is my observation **BIG**?	YES	NO
Does it **show details**?	YES	NO
Does it use **correct colors**?	YES	NO
Is it **fully colored**? (no white space)	YES	NO
Does it have **labels**?	YES	NO
Is it **neat and organized**?	YES	NO
I can improve by …		

Chapter 5

Figure 5.17. Student observational drawing with checklist.

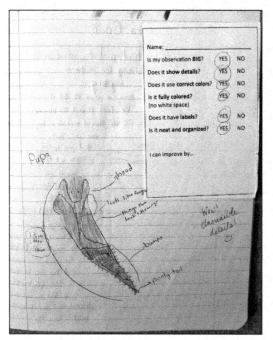

a checklist that students glued into their notebooks next to their drawings to help them reflect on their work (Figure 5.17). Students were excited to compare their initial drawings with their new ones and were quite proud of their improvements. The checklist became a tool that I continued to offer students throughout the year as needed. While this specific workshop took place with a class of third-grade students, it could be used with any elementary grade with simple modifications for the age level of the students involved.

Claim and Evidence Workshop

My second notebook workshop example also comes from my third-grade class, although it is one that I have used in all my upper elementary classes at one point or another. This particular instance was during student investigations into the effects of forces on objects (3-PS2-1: Plan and conduct an investigation to provide evidence of the effects of balanced and unbalanced forces on the motion of an object). As I reviewed students' notebook entries, I noticed that many students were not citing experimental data to support their claims and were instead writing general statements such as "I did an experiment." I knew it was time to pause for a notebook workshop on writing strong evidence statements. Before the next class, I selected three students' work to share anonymously, ranging from a very general statement to a very strong and specific statement including data.

In the next class, I set the purpose for our workshop: identifying and writing strong evidence statements to support our claims. I showed students the three sentences from their classmates' work and asked them to discuss which statement provided the strongest support for the claim. Students immediately and unanimously identified the strongest statement, so I then challenged them to analyze that statement further to figure out why it was the strongest. As a class, we determined that the inclusion of experimental data was necessary for a compelling piece of evidence. At this point, I released students to return to their notebooks and revise their evidence statements as I circulated to provide individual assistance.

Other Notebook Workshops

While these two examples focus on specific skills, I also use notebook workshops for broader concepts. Sometimes, I highlight a specific way a student has organized their notebook entry and ask them to share their work with the class using the document camera, inviting the other students to try the approach themselves. Other times, our workshop focuses on an effective way to record data. I don't go into a school year with a typical list of notebook workshop topics in mind (although some, like the previous examples, tend to be necessary each year); instead, I plan these as needs emerge from my students' work. Doing this allows me to meet students where they are and move them forward, and it also allows me to recognize and highlight individual students' strengths as they are revealed in their work. Although I could present the same workshops before students ever create an observational drawing or write a claim and evidence statement, I've found that waiting for them to try these skills first and then conducting a workshop with their peers' work is much more powerful. Students are more likely to take the principles to heart and adopt them in their own work when a friend or peer has demonstrated them first. In this case, imitation is the sincerest form of flattery—and a terrific learning tool!

Even though I prefer to have students create their own notebook entries whenever possible, there are times when I break my own rule. Often, these instances are due to situational constraints that outweigh the value of a student-created data table or organizer. For example, my fifth-grade students spent an entire week at our outdoor campus this past year, engaged in various fieldwork techniques such as stream quality monitoring, tree identification, and finding and identifying soil macroinvertebrates. Each of these experiences involved the creation of tally charts and data tables, and my fifth-grade students were more than capable of creating their own at that point in their science education. However, I chose to provide these elements for students to glue into their notebooks in the interest of saving time that could be spent doing actual fieldwork. In this particular case, our time was better spent finding and identifying organisms than practicing making a data table.

In this chapter, I've described the teacher-directed ways in which I have my students set up and organize their notebooks as well as the student-centered approach I use for their notebook entries. I've also explained how I have adapted literacy models and practices like the gradual release of responsibility model and writing workshop to enhance my students' use of their notebooks. In the next chapter, I share specific instructional strategies aligned with the science and engineering practices and crosscutting concepts, all of which can be placed on this teacher-directed to student-centered continuum.

References

Calkins, L. 2006. *A guide to the writing workshop, grades 3–5*. Portsmouth, NH: First Hand.

Collins, A., J. S. Brown, and A. Holum. 1991. Cognitive apprenticeship: Making thinking visible. *American Educator* 15 (3): 6-11.

Collins, A., J. S. Brown, and S. E. Newman. 1989. Cognitive apprenticeship teaching and the crafts of reading, writing, and mathematics. In *Knowing, learning, and instruction*, ed. L. B. Resnick, 453–494. Hillsdale, NJ: Lawrence Erlbaum Associates.

Pearson, P. D., and M. Gallagher. 1983. The instruction of reading comprehension. *Contemporary Educational Psychology* 8 (3): 317–344.

Spencer, H. 1854. The art of education. *North British Review* 41: 152. Edinburgh, Scotland: W. P. Kennedy.

CHAPTER 6

Using the Notebook as a Thinking and Learning Tool

The fun in science lies not in discovering facts, but in discovering new ways of thinking about them.

—Sir Lawrence Bragg, *A Short History of Science*

Science notebooks are a safe place for students to record observations, wrestle with new ideas, and synthesize information from multiple activities into a coherent understanding of a concept. As described in Chapter 5, I seek a balance between teacher-directed and student-directed entries. In certain situations, I give specific directions for a notebook entry, often paired with a graphic organizer or series of questions to be answered. At other times, I give students a goal (for example, "Record your observations of your mealworm in a sketch and description") and let them decide exactly what form that entry will take. The specificity of my directions depends on my learning objectives for the activity, the time of year in which the activity is taking place, the age and skill level of the students, and whether I want students to arrive at the same conclusion or bring a variety of ideas to our scientist meeting. As a result, my students' notebooks are a mix of specific strategies and freely constructed entries. In this chapter, I share some specific strategies and notebook entries that support the eight science and engineering practices as well as guide students to use the crosscutting concepts as a lens through which to view phenomena. Far from being an exhaustive list, this is only a sampling of ways that notebooks can support three-dimensional teaching and learning.

Trying to organize strategies into neatly defined categories based on the science and engineering practices quickly became overwhelming and resulted in a long, unwieldy list. It was also frustrating to try to limit a specific strategy or entry to a single SEP when many, if not all, addressed several of them. Turning to the literature, I discovered that researchers, professional development providers, and teachers had the same experience when working with eight distinct practices (McNeill, Katsh-Singer, and Pelletier 2015). In response, McNeill and colleagues grouped the practices into three categories: investigating practices, sensemaking practices, and critiquing practices. The authors note that such a grouping is an oversimplification of these dynamic practices and that other groupings are possible, and I have moved practice 5 (using mathematical and computational thinking) to the sensemaking practice group instead of the investigating group. In my classroom, students engage in practice 5 most frequently during sensemaking activities. With this one change, the framework provided the clarity and language I needed to organize and discuss the myriad of ways in which I use science notebooks in my teaching. Table 6.1 shows these modified groupings, along with an overview of aligned notebook entries for each group.

Table 6.1. The eight SEPs grouped into investigating, sensemaking, and critiquing practices, along with aligned notebook entries.

	Investigating practices	**Sensemaking practices**	**Critiquing practices**
Science and engineering practices	1. Asking questions and defining problems 3. Planning and carrying out investigations	2. Developing and using models 4. Analyzing and interpreting data **5. Using mathematical and computational thinking** 6. Constructing explanations	7. Engaging in argument from evidence 8. Obtaining, evaluating, and communicating information
Aligned notebook entries	• Observational drawings • See/Think/Wonder • Notice & Wonder • Question Formulation Technique (QFT) • Photographs of students using tools and conducting experiments • Drawings and summaries of experimental setup • Written descriptions of experimental procedures	• Documentation of qualitative data • Documentation of quantitative data • Graphs • Photographs of and reflections on collaborative whiteboard models • Diagrams with explanations • Revised models • Engineering design models • Summary tables • Big idea pages • Reflective paragraphs	• Claim and evidence T-charts • Formal written arguments • Sketchnotes • Other T-charts • Research notes • Storyboards and rough drafts

Source: Adapted with permission from McNeil, Katsh-Singer, and Pelletier (2015).

Note: Boldface indicates the practice that has been moved to a different grouping than in the original source.

Although I prefer traditional paper notebooks, almost all the strategies included in this chapter can be easily translated to a digital format. The Google suite of applications, including Google Docs, Drawings, Slides, Sheets, and Jamboard, offers student-friendly options to paper-and-pencil work, and other applications such as Flipgrid and voice recording software provide alternatives to writing or typing at all. While the examples included throughout this chapter were all created with paper and pencil, it is the strategy and not the medium that I want to highlight.

Figure 6.1. Observational drawings by elementary students.

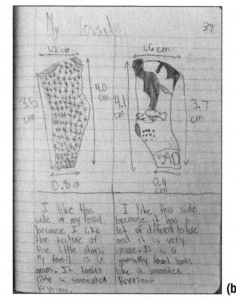

(a)   (b)

A first-grade student's observational drawings of plant parts (a); and a fourth-grade student's observational drawings of a fossil collected on a field trip (b).

Notebook Entries for Investigating

As students investigate, they pose questions, plan experiments, and then conduct experiments using scientific tools. As with scientists, students' questions are frequently sparked by an observation of a phenomenon. I've found that elementary students often need prompting to observe deeply enough to ask thoughtful questions, and observational drawing and description is an effective strategy for doing so. In this section, I share notebook entries for investigation, including observational drawings and descriptions, strategies to help students ask the right questions, photographs of students using tools and conducting experiments, drawings and summaries of experimental setups, and written descriptions of experimental procedures.

Observational Drawings and Descriptions

Students in my classes engage in a fair amount of observational drawing and description each year. Why? Aside from the obvious curricular connections, I've found that by honing their observation skills, students become more curious and pose more (and better) questions that deepen our investigations. Observational drawing is emphasized in our preschool and kindergarten's Reggio Emilia program as well, so my elementary students come to me with more highly developed artistic skills than they might otherwise have. Nevertheless, work remains to be done to improve the quality of their artistic documentation and written descriptions.

Figure 6.1 shows two observational drawings done by our elementary students: parts of a plant (6.1a) and a fossil of an ancient animal (6.1b). As these photographs illustrate, I tend to give

Figure 6.2. A fifth-grade student's owl pellet observation (pre-dissection) with checklist of required elements.

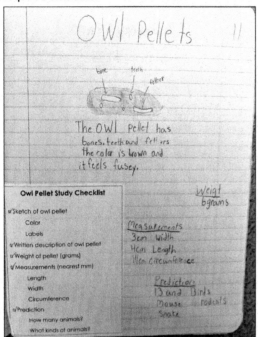

Figure 6.3. Student observations of Wisconsin Fast Plants.

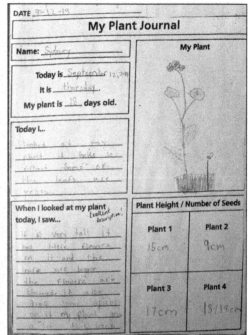

students a great deal of choice in how they organize their sketches and descriptions on the page. I often give a list of details that they must include in their observations, usually in the form of a checklist, as shown in the owl pellet observation in Figure 6.2. However, at times, I rely on graphic organizers to guide students in their observations, such as my students' observations of Wisconsin Fast Plants throughout their life cycle. The graphic organizer is included in the kit, and I've found that it provides the needed structure and scaffolding to support this work. As shown in Figure 6.3, students are asked to include more information in their observations than they might if left to their own devices.

I use a notebook workshop approach (see Chapter 5) to help students improve the quality of their observational drawings. Another instructional point that often arises is the need to use objective language in written descriptions. Students frequently use subjective language to describe what they are observing, including words like *pretty, beautiful, weird, cool, gross,* and *disgusting.* When I notice such words in notebook entries, I first discuss them with individual students, and then I make time for a whole-class workshop session as soon as I can. Explaining the difference between subjective and objective language (I use the terms *opinion words* and *fact words* with students) and the importance of using objective language in scientific observations supports students in being as precise as possible in their descriptions of objects. Figure 6.4 shows feedback I might leave for a student who is using subjective language in her descriptions.

Figure 6.4. Feedback for a student regarding the need to use objective language in descriptions.

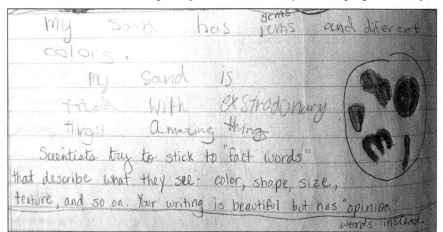

Strategies to Help Students Ask the Right Questions

While elementary students are excellent at posing questions, they need instruction and scaffolding to ensure that their questions are ones that can be answered by the collection of data. Strategies that support students in posing questions, linking observations to questions, and determining whether a question is testable include See/Think/Wonder, Notice & Wonder, and the Question Formulation Technique (QFT).

See/Think/Wonder

See/Think/Wonder is a thinking routine developed by Project Zero (2019), a research center at the Harvard Graduation School of Education. It invites students to observe carefully, interpret observations, and pose questions for further investigation. Though this thinking routine can be used across the curriculum, I have found it to be an invaluable strategy for helping students respond to scientific phenomena and find an entry point into an investigation. Many of my units of study begin with a See/Think/Wonder that students continually revisit as the investigations unfold.

The See/Think/Wonder routine is easy to implement with students. First, determine how students will experience the phenomenon or problem. Possibilities include, but are not limited to, a video clip, set of natural objects, sequence of images, or graph. Give students time to observe carefully, and then lead them through this series of three questions:

- What do you see?
- What do you think about what you see?
- What does it make you wonder about?

Ideally, students answer these three questions in succession, linking their observations, interpretations, and questions. However, I've found that many students prefer to tackle one question at a time, first recording all their observations, then their interpretations, and finally their

Chapter 6

Figure 6.5. A See/Think/Wonder chart completed as students examine different maps showing Earth's physical features.

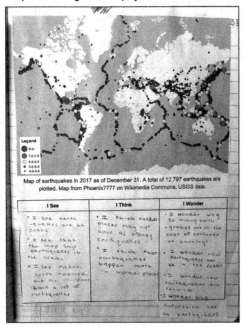

Map of earthquakes in 2017 as of December 31. A total of 12,797 earthquakes are plotted. Map from Phoenix7777 on Wikimedia Commons. USGS data.

questions. Additionally, elementary students don't always have an interpretation or a question about every observation they make. I want this thinking routine to produce the widest array of observations, ideas, and questions possible, so I encourage students to use the routine in the way that makes the most sense to them. Follow-up discussions can then help them see the connection among these three components.

See/Think/Wonder works well as a whole-class discussion, and I often use this approach with my very young students, for whom the demands of writing would impede their thinking. With older students, I prefer to have them complete individual See/Think/Wonder charts in their notebooks before compiling our ideas in a class discussion. (A graphic organizer that I have used to introduce students to this process is included in the appendix on p. 119. However, students can easily create this simple three-column organizer in their notebooks.)

I have used this thinking routine extensively in my classroom and in many different scientific contexts. For example, in an Earth science unit of study, students investigate the locations of Earth's features by studying a variety of maps, including raised-relief maps and maps that show the locations of earthquakes and volcanoes. As they rotate through stations with these different maps, they complete See/Think/Wonder charts in their notebooks, as shown in Figure 6.5. This thinking routine allows students to identify patterns in the data shown on the different maps, helping them move toward meeting that performance expectation. After students have completed their individual thinking routines, we gather as a class to share and compare observations, ideas, and questions. What emerges during our scientist meeting frames how the rest of this unit of study will progress.

Another example of a See/Think/Wonder completed in students' notebooks (Figure 6.6) is from a third-grade study of weather. At the beginning of the unit, I chose to use the See/Think/Wonder thinking routine to elicit students' prior knowledge and questions using a concept they encountered frequently in their daily lives: a weather forecast. Discussion with students revealed that their experiences with forecasts were limited to announcements on the news or radio or the weather app on their parents' phones. I presented students with a printed copy of the 10-day forecast for our area (6.6a) and asked them to engage in the thinking routine around the forecast. Their ideas and questions (6.6b) were an important source of information for planning the unit.

A final example of a See/Think/Wonder thinking routine is from a multigrade project conducted at the beginning of the school year (Figure 6.7). At the end of the previous year, I had asked that students whose families were traveling to a beach on vacation each collect a small sample of sand and bring it back to school in a labeled container. I had a fantastic response, with samples from

Figure 6.6. See/Think/Wonder completed in third-grade students' notebooks from a study of weather.

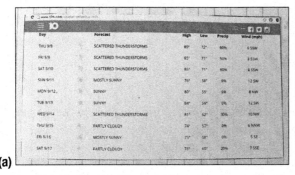
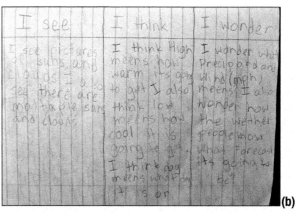

(a)   (b)

Printed 10-day local forecast for student use in a See/Think/Wonder thinking routine (a); and a student's completed See/Think/Wonder chart in response to the 10-day forecast (b).

all over the United States and even a few international ones. In August, students observed samples from different locations using stereomicroscopes and practiced their observational drawing and description skills. In addition, I created a bulletin board mapping the locations of the samples along with a small amount of sand from each location (6.7a). Students took time to explore the display and completed See/Think/Wonder charts in their notebooks (6.7b). Their observations of similarities, differences, and patterns in the colors and texture of sand launched the class into an exploration of weathering, erosion, and deposition (as appropriate for each grade level).

See/Think/Wonder is a simple strategy, yet it can produce powerful results. I have learned through experience that students need repeated modeling, discussion, and practice to become

Figure 6.7. See/Think/Wonder completed in students' notebooks from a study of sand samples.

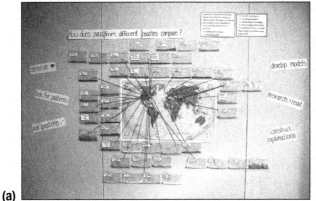
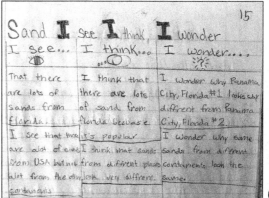

(a)   (b)

Map of schoolwide sand samples collected over the summer break (a); and a student's completed See/Think/Wonder chart in response to the sand study map (b).

fully proficient with it. In many cases, students' first attempts at this thinking routine are quite literal and surface level. Don't be discouraged. Discuss their ideas and give your own examples to show higher-level thinking, particularly in the "Think" section. Also, be sure that the phenomenon you have selected is rich enough to provoke deep reflection and questioning. Sometimes a lack of high-quality responses on the students' part is an indicator that the subject of the routine is too simple or too complex for them to fully engage.

Notice & Wonder

A similar activity to See/Think/Wonder, Notice & Wonder helps students observe deeply and turn their observations into related questions. I use this strategy often when students are observing living organisms, fossils, or rocks. Figure 6.8 shows two examples of Notice & Wonder charts, which show how each question is directly connected to an observation: a student's observations and questions about the Wisconsin Fast Plants (6.8a) and a student's chart following her observation of darkling beetles as part of an investigation into mealworm life cycles (6.8b). I have found that this questioning activity is most successful after students have spent a good amount of time immersed in observations and possibly observational drawing and description. They are able to draw on the rich details captured in their observations and turn those into questions that can guide further investigation. I have also found that this graphic organizer is incredibly helpful for students, especially when they are new to the activity. The example and the arrows connecting the observation and question rows really help students understand the link between these two pieces of information. Without the arrows, students tend to write lists of unrelated observations and questions. This is a case where I tend to use the organizer longer, turning over responsibility to students only when they have demonstrated that they understand the purpose of the activity. Finally, you will notice in comparing the two examples that some students repeatedly rely on one or two question starters (e.g., *how* and *why* in 6.8a), while others generate a more varied list of questions (6.8b). This is another example of a notebook entry serving as formative assessment, and I work with students to help them develop a more sophisticated and diverse set of questions.

Figure 6.8. Completed Notice & Wonder charts with student observations and questions.

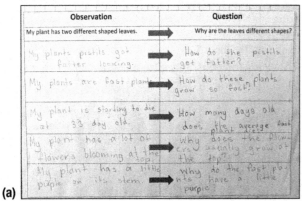

(a)

(b)

A Notice & Wonder chart following observation of Wisconsin Fast Plants (a); and one following observation of a darkling beetle (b).

Question Formulation Technique

The Question Formulation Technique (QFT) was created by the Right Question Institute as a technique to empower parents to ask questions during parent-teacher conferences. Since its creation, the technique has been adopted for a wide variety of purposes, including teaching students to ask questions. A simple yet powerful strategy, the QFT stimulates three different types of thinking: divergent, convergent, and metacognitive. The QFT centers on a stimulus, called a QFocus, and then follows four steps: produce questions, improve questions, prioritize questions, and discuss next steps. The institute has produced many helpful resources and videos with regard to the QFT, all available on its website (see QR Code 6.1).

QR Code 6.1. Introduction to the Question Formulation Technique on the Right Question Institute's website, with links to resources and more information.

While the QFT can certainly be used in science class in its original form, I use a variation to help my students learn to differentiate between scientific (testable) and nonscientific (nontestable) questions. The initial steps of the process are the same as in the original QFT: students engage with a QFocus (in this case, a phenomenon) and generate as many questions as possible, which they can jot down in their notebooks (Figure 6.9). The "improve questions" step is where this deviates from the original protocol. Instead of classifying questions as open or closed, students sort questions into several categories: observational (can be answered with the senses), research (can be answered by reading, watching a video, or consulting an expert), or testable (can be answered through investigation). I like to have students write questions on sticky notes and sort them physically (Figure 6.10), but this sorting could be done in a written notebook entry. The final steps, prioritize questions and discuss next steps, remain the same as in the original.

The revised QFT strategy is well suited for three-dimensional learning, as it can help students pose questions after engaging with an anchoring phenomenon. The prioritized testable

Figure 6.9. A student generates questions during the first step of a QFT.

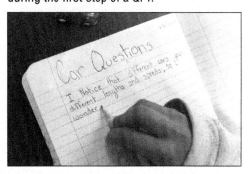

Figure 6.10. Sorting questions into categories in a revised QFT.

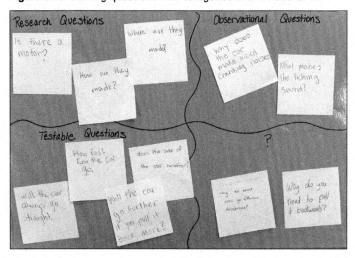

Chapter 6

questions can then serve as the basis for subsequent investigations as students work to understand, model, and explain the phenomenon. Questions generated through the QFT could form the basis of a driving question board or class question board that organizes students' questions and provides a road map for learning (Phenomenal Science Team 2017). Once a question has been identified, students need to determine how to go about answering it. This begins with simple investigations based on the concept of a "fair test" in the primary grades and builds to controlling variables in the intermediate grades. Embedded in this practice are the observation, measurement, and procedural skills needed for students to successfully collect data from an investigation. Fortunately, science notebook entries can support the development of these underlying skills as well as the overarching SEP.

Figure 6.11. A second-grade student's notebook entry showing the proper use of a graduated cylinder.

Photographs of Students Using Tools and Conducting Experiments

I give full credit for this brilliant idea to my colleague and friend Karen Scranton, who has also taught our youngest elementary students science. When teaching them how to use a scientific tool, she often takes pictures of the students using the tool correctly. Students glue these pictures of themselves in their notebooks and sometimes add a reflection about the process. For example, Figure 6.11 shows second-grade students demonstrating how to use a graduated cylinder to measure the volume of a liquid, and Figure 6.12 shows a student and her partner carrying out an investigation. Adding photographs of students at work to notebook entries can promote reflection and metacognitive thinking, and the photos serve as a great reminder of proper technique in the future.

Figure 6.12. Two students work together to conduct an experiment.

Drawings and Summaries of Experimental Setups

Students sometimes draw pictures illustrating their experimental setups and write short descriptions of their procedures. This is true for both teacher-directed investigations (Figure 6.13) and student-directed investigations (Figures 6.14, p. 67, and 6.15, p. 68). Figure 6.13 shows a student's choice to recap a demonstration I performed for the class to help answer students' questions about whether air is matter.

Figure 6.14 shows a second-grade student's initial plan (6.14a) to investigate what plants need to survive and a teacher-created graphic organizer (6.14b) that students completed as a whole class while they collaboratively planned their experiment. My colleague Karen used questions to help students begin to more explicitly consider the elements of a fair test, although this particular student's work shows that she understandably needs further practice in this area.

Figure 6.15 (p. 68) shows an example of a third-grade student's plan for a collaborative investigation into factors affecting a car's movement down a ramp. This was a rather open-ended assignment in which I gave students the materials and tasked them with planning and conducting an experiment in collaborative groups. Analyzing students' notebook entries gave me valuable information about the areas of experimental design in which students needed support. This student used a sketch to show the experimental setup and recorded two data points (likely from the same trial), without further explanation

Figure 6.13. A student's sketch and summary of a demonstration that air takes up space.

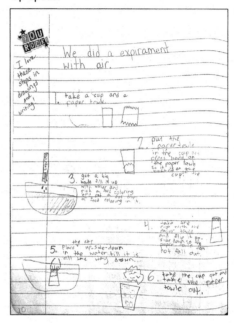

Figure 6.14. Planning process for a student investigation.

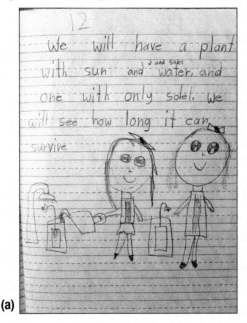
(a)

(b)

A second-grade student's original plan for an investigation (a); and a teacher-created graphic organizer to refine the planning process (b).

Figure 6.15. Sketch and plan for a student-designed investigation of factors affecting a car's motion down a ramp.

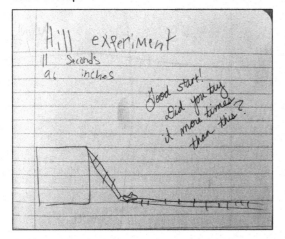

Figure 6.16. Experimental planning graphic organizer as completed by a fourth-grade student.

Variable	How far we
What one thing will be changed in your experiment?	pull the car back.
Controls	same surface, starting place, and car.
What will you keep the same during your experiment?	
Data	How far the car goes in CM.
What will you observe or measure to answer your question?	

or context. In my experience, third-grade students are developing their understanding of a "fair test" and typically do not include all relevant information in their descriptions of experimental procedures, making their sketches even more important in conveying information. This type of write-up paves the way for more formally written experimental procedures.

Written Descriptions of Experimental Procedures

Students in the intermediate grades are ready to more formally plan and describe an experiment, but I have found that students need more support with identifying the important elements of the experiment (variable, controls, data) than with writing the steps, so I usually provide a simple graphic organizer (Figure 6.16) to help them think through these questions.

After students have successfully answered the questions, they are typically ready to write a procedure without the use of a graphic organizer.

Notebook Entries for Sensemaking

Students engaged in sensemaking use a large number of science practices, including interpreting and analyzing both qualitative and quantitative data, using mathematical and computational thinking, developing and using models, and constructing explanations. These notebook entries may include graphs, models, diagrams, and other aids to support student thinking.

Representations of Qualitative Data

In addition to observations of natural objects, some investigations involve the collection of qualitative data. In these cases, students best represent the data through diagrams and writing. Qualitative data is especially suitable for primary students who are still developing basic number sense and arithmetic skills and can be easily depicted in open-ended notebook entries. Figure 6.17 shows two examples of student data based on their observations. One is an entry in a first-grade student's notebook recording data from her daily observation of wet and dry corn kernels (6.17a) as her class works through the Next Generation Science Storyline unit Why Is Our Corn Changing? (Next Generation Science Storylines Team 2019). The other is an entry from another first-grade

Figure 6.17. Entries from first-grade students' notebooks recording qualitative data from their observations.

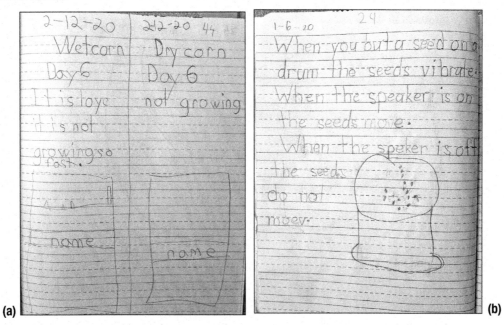

(a)   **(b)**

Data recorded from observations of dry and wet corn kernels on Day 6 of an investigation (a); and a drawing and an explanation of seeds moving on a drum when a speaker playing music is placed nearby (b).

student's notebook (6.17b), depicting her observations of seeds placed on a drum with and without a speaker playing music nearby. Although some quantitative data is collected in primary-level investigations, drawings, observations, and descriptions play a large role.

Qualitative data is also valuable in the intermediate grades. While students are more capable of collecting and analyzing numerical data, some investigations do not lend themselves well to measurement. Figure 6.18 (p. 70) shows two examples of qualitative data recorded by students in intermediate grades. One is a notebook entry from a fourth-grade student (6.18a) using simple sketches and writing to document her qualitative observations of energy transfer between colliding marbles. The other documents a student's observations of the changes in a stream table as a result of erosion (6.18b). Continuing to practice documenting observations in sketches and written descriptions is worthwhile, even as students begin to collect more quantitative data.

Representations of Quantitative Data

Students in the intermediate grades increasingly rely on quantitative data, including measurements of temperature, time, mass, and distance traveled. Science notebooks serve as a physical location for recording data as well as a learning opportunity in terms of how best to organize data from an investigation.

Figure 6.18. Entries from notebooks showing qualitative data recorded by students in intermediate grades.

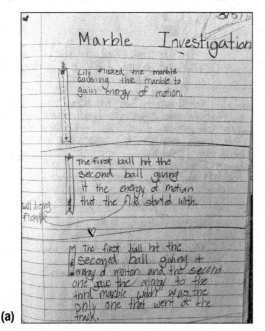

(a)

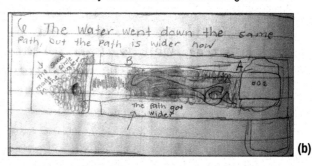

(b)

A fourth-grade student's observations of the transfer of energy in colliding marbles (a); and a student's observations of erosion in a stream table (b).

Figure 6.19. A student created this data table with teacher guidance to investigate the question "Which seeds are best adapted to landing in the sun?" A check mark indicates which seed shapes landed in the sun and an *X* indicates which landed in the shade.

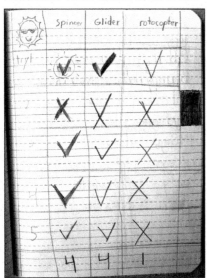

I try to balance providing appropriate amounts of scaffolding for students with allowing them to try out different strategies and approaches for organizing information in their notebooks. This is particularly true for data collection. We begin using tally charts and data tables in whole-class activities in the primary grades, and I sometimes assist students in creating simple data tables that can be completed with a check mark, as shown in Figure 6.19. While these data tables don't actually include numerical data, they do introduce the concept to students in a developmentally appropriate manner.

As students progress through the elementary years, they increasingly work with quantitative data and become more proficient at using data tables (Figure 6.20). Students in third grade begin working with more quantitative data through their studies of topics such as weather, forces, and traits. I frequently supply these students with premade tally charts and data tables (6.20a) and take time for direct instruction, discussion of the components of a data table (columns and rows), and modeling how to enter data into these organizers correctly. For students in fourth grade, I typically provide one or two data tables but quickly move to drawing a data table on the board and asking students to copy it (6.20b), eventually progressing to giving verbal instructions ("Draw a table with four columns and three rows") by the end of the year. By fifth grade, students are becoming proficient in creating simple data tables, a skill that is also taught and practiced in our math classes.

Graphs

Creating graphical representations of data, including pictographs and bar graphs, follows a similar progression through the elementary years (Figure 6.21). Students in the primary grades gain experience with graphs through whole-class experiences, while students in the intermediate grades become increasingly independent in their graphing skills. As with other types of notebook entries, I provide greater scaffolding and support in the earlier grades, typically by supplying predrawn axes with labels (6.21a). Students continue practicing, aided by a graphing checklist (6.21b), with the goal of independently creating bar graphs by the end of fifth grade (6.21c).

Figure 6.20. Examples of data tables used by students at different grade levels.

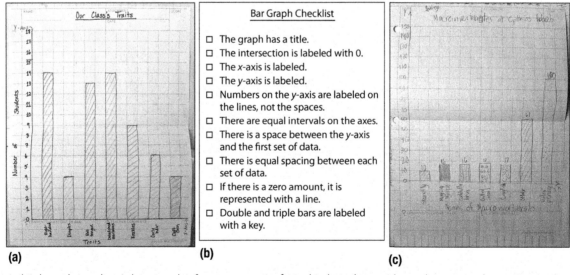

(a) **(b)**

A teacher-created data table for third-grade students to track hourly temperatures (a); and a data table created by a fourth-grade student with teacher support (b).

Figure 6.21. Progression of graphical representations of data through the elementary grade levels.

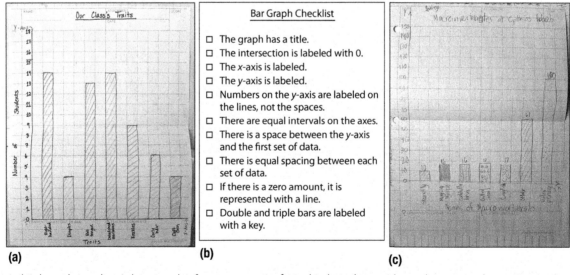

Bar Graph Checklist

- ☐ The graph has a title.
- ☐ The intersection is labeled with 0.
- ☐ The x-axis is labeled.
- ☐ The y-axis is labeled.
- ☐ Numbers on the y-axis are labeled on the lines, not the spaces.
- ☐ There are equal intervals on the axes.
- ☐ There is a space between the y-axis and the first set of data.
- ☐ There is equal spacing between each set of data.
- ☐ If there is a zero amount, it is represented with a line.
- ☐ Double and triple bars are labeled with a key.

(a) **(b)** **(c)**

A third-grade student's bar graph of common traits found in her class, with teacher-created axes, interval, and labels (a); a bar-graphing checklist used by students in fourth and fifth grades to self-assess their work (b); and a bar graph created independently by a fifth-grade student (c).

Chapter 6

Working With Data

In the elementary years, there is quite a bit of overlap between analyzing and interpreting data and using mathematical and computational thinking. The reason for this overlap is simple: Students have not yet developed the mathematical skills to analyze data in sophisticated ways. That being said, there are a few instances in which students in the intermediate grades do begin to work with small data sets, such as identifying the median of a data set or calculating the mean. Whenever possible, I try to present situations in which students think critically about what these measures of center actually mean in terms of their data. For example, as we discuss the importance of conducting multiple trials (and then do just that), I ask students which trial's data point should be used as evidence. Students typically identify the median as the most logical choice, presenting a teachable moment for introducing that mathematical term. Figure 6.22 shows two examples of students' use of the concept of the median. One reveals a third-grade student's thought process as the student compares data from multiple trials and unknowingly settles on the median as the most accurate measurement (6.22a). In the other, a fourth-grade student identifies the median data points by circling them in her data table (6.22b). Although these concepts are taught in middle school according to the *Common Core State Standards for Mathematics* (NGAC and CCSSO 2010), they are introduced in my independent school's mathematics curriculum in the upper elementary grades. I have also found that students can more easily grasp the concept of a median or mean when working with data that they have collected.

Figure 6.22. Examples of students' work reflecting their use of the concept of a median.

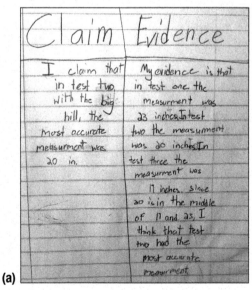

(a)

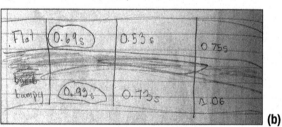

(b)

A third-grade student intuitively uses the mathematical concept of a median as she provides evidence to support her claim (a); and a fourth-grade student identifies the median data points by circling them in her data table (b).

In a similar way, I have asked fourth-grade students to use the class's Fast Plant data to predict the height that a plant will be on a certain day of its life. Students bring a range of creative ideas to this challenge but often end up using the mode or median without knowing it. Figure 6.23

shows one student's attempt to predict the height of a plant based on her class's experimental data. While her strategies may not have led to the correct answer, this exercise gave an opportunity to teach various measures of center, including mean, median, and mode. This challenge presents an opportunity to introduce the concepts of mode and median in a meaningful way and introduce any other measures of center that were not brought up in discussion.

Models

Students use the data they've collected, analyzed, and interpreted to develop and revise models. Models can take many different forms, including diagrams, drawings, physical replicas, mathematical representations, analogies, and computer simulations. As with other practices, modeling becomes increasingly sophisticated as students mature. Primary students represent concrete events or design solutions with diagrams, drawings, physical replicas, dioramas, dramatizations, or storyboards. In the intermediate grades, students not only continue to construct various models of increasing sophistication but also revise their initial models and use models to make predictions. While all models necessarily contain parts and relationships between parts, there are different ways to

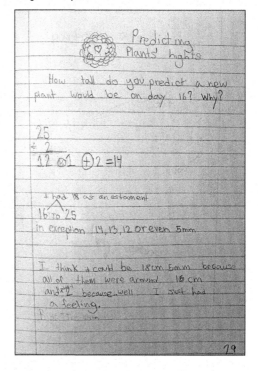

Figure 6.23. A student uses data to reason creatively as she attempts to predict the height of a plant.

scaffold model creation within a science notebook. I've used various scaffolds with my students, including photographs of whiteboard models, diagrams with explanations, individual models with opportunities for revision, and engineering design models.

Photographs of Whiteboard Models

Many teachers have students use whiteboards to create models. Whiteboarding has several advantages for this science and engineering practice: It allows for collaboration, feels less intimidating than working on paper because students can more readily erase their work and start over, and makes it easy for students to revisit and revise their work. Its disadvantage? It is only temporary—whatever thinking students record can't be preserved.

My solution? I photograph completed models on whiteboards and print copies for students to glue into their notebooks. In addition, I ask students to explain their models in writing along with the pictures. This provides an important source of formative assessment, especially when students have worked collaboratively on a model. I can discern students' individual levels of understanding of the group model by reading their writing. It also gives students an opportunity to include additional parts (components) or represent relationships between parts of the model that the entire group may not have agreed on but that an individual student believes are important. These

Figure 6.24. Collaborative whiteboard model and student explanation of how sand is formed.

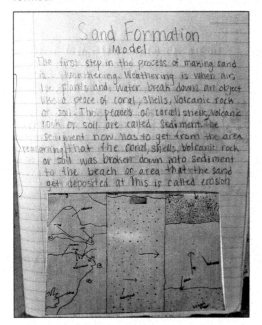

additions may be crucial for explaining the phenomenon or could turn out to be extraneous.

Figure 6.24 shows an example of this type of notebook entry. Students created whiteboard models in collaborative groups as part of their start-of-year investigation of sand brought back from different vacation locations around the world. They created an initial model to show their thinking about how sand was formed, and then they revised and finalized their models after further investigation and reading about weathering, erosion, and deposition. Students glued a picture of their team's model into their notebooks and described what was shown in their model.

Diagrams and Descriptions

Another way I've used science notebooks to scaffold modeling is by asking students to create a diagram in their notebook and then describe what they observed. One example is a first-grade student's model of how light from a flashlight travels through a piece of plastic tubing (Figure 6.25a). The diagram shows that light travels all the way through the tube when it is straight but only partially through when the tube is bent. Her accompanying description of what she observed states that "the curvy one is hard to see … the round one is easy to see." This notebook entry is a rich stepping-stone toward helping students see the explanatory or predictive power of a model. Follow-up questions to this student might include "Why do you think the light didn't travel all the way through the tube? What if I had a different tube that was curved? Do you think there's a time when light can travel around a corner?"

A second example shows a fourth-grade student's diagram and observations of a simple circuit (Figure 6.25b) from an investigation of the transfers and transformations of energy. The student has drawn and labeled a diagram of the circuits and uses her current understanding of energy to describe what she saw. To help this student connect evidence to explain why the light goes on, you might ask, "Why do you think the battery is causing the light to go on? How could you represent that in your model?" Identifying the cause-and-effect relationship is a developmentally appropriate expectation for a fourth-grade student.

Individual Models With Opportunities for Revision

Although we often develop models collectively, there is great value in providing the time and space for students to own their thinking. One such example is a fifth-grade student's model explaining how Yellowstone National Park's wildlife populations might have changed after the reintroduction of wolves in 1995 (Figure 6.26). In such models, students use multiple modalities (text, drawing, symbols, and in some cases, color) to represent their thinking about the phenomenon in question,

Figure 6.25. Examples of diagrams with accompanying explanations.

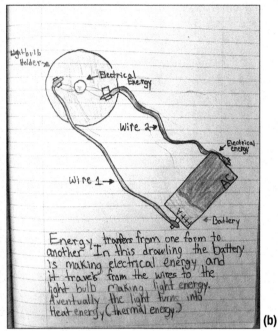

(a) (b)

A first-grade student's diagram showing how light travels through a straight and curved tube (a); and a fourth-grade student's diagram of a circuit and explanation of how energy is transferred and transformed (b).

which is an excellent opportunity for student choice. In this case, a graphic organizer provided a second layer of scaffolding, supporting students as they modeled changes occurring over time.

Students also have the opportunity to revisit their work throughout a unit to add to, remove, or otherwise revise their models as their thinking changes, creating a powerful record of their emerging understanding. While this particular student made all her changes in pencil, asking students to use different colors or otherwise notate revisions can help track their progress in understanding how the phenomenon occurred over time.

Figure 6.26. A fifth-grade student's model depicting Yellowstone National Park before, during, and after the reintroduction of wolves.

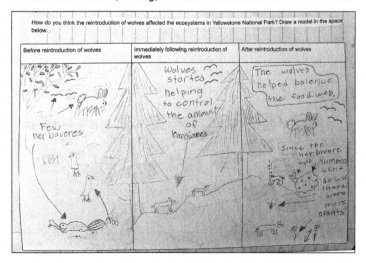

Chapter 6

Figure 6.27. A student's individual designs for a hand pollination device.

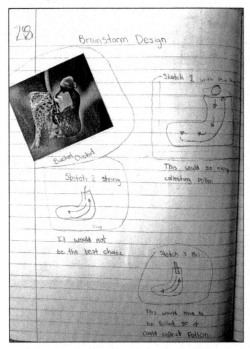

Engineering Design Models

Models can also be created as students generate possible solutions to a problem (design challenge). In Figure 6.27, a student has drawn models of different designs for a hand pollinator as part of a larger study on pollination and the Engineering Is Elementary "The Best of Bugs: Designing Hand Pollinators" module (Museum of Science 2020). In this case, students sketched individual ideas for devices before comparing their designs with those of their partners. For their group design, students might choose to include common features, a feature unique to one student's design that they think will allow the group design to meet the criteria for success, or both.

Explanations

A major goal of the investigations taking place in a science classroom is that students are able to use the evidence they've gathered to construct an explanation for a phenomenon or design a solution to a problem. Explaining how or why something happens or the relationship between two things draws on several of the preceding science and engineering practices, such as developing and using models. Strategies for supporting this type of work in science notebooks include summary tables, big idea pages, and reflective paragraphs.

Summary Tables

A summary table is an incredibly helpful tool in units that span a considerable length of time, as it affords students with a record of what they have done and read. Though many variations exist, a summary table typically consists of several columns detailing the activity, observations and patterns, and connections to the essential question or anchoring phenomenon. Students provide the ideas to complete a row of the table after each activity, and as the unit unfolds, they develop an increasingly complete understanding of the phenomenon. For more information on summary tables, see the Ambitious Science Teaching website, which offers explanations, examples, and tips from teachers who use them successfully in their classrooms (see QR Code 6.2).

QR Code 6.2. Face-to-face tools, including summary tables, from Ambitious Science Teaching Development Group.

I have used summary tables with my classes for several years now as a whole-group activity but wondered if there would be value in guiding students to create personal summary tables in their notebooks. I decided to try this approach with my fourth-grade students during our study of energy and modeled how to create a summary table in their notebooks, as shown in Figure 6.28. This individualized summary table was a helpful resource for students to refer to, and it provided me with an additional source of formative assessment data in terms of how students were conceptualizing each investigation within the unit.

Figure 6.28. A student's summary table from a unit on energy. Note that the term *potential energy* is officially introduced in grades 6–8 in the *NGSS*.

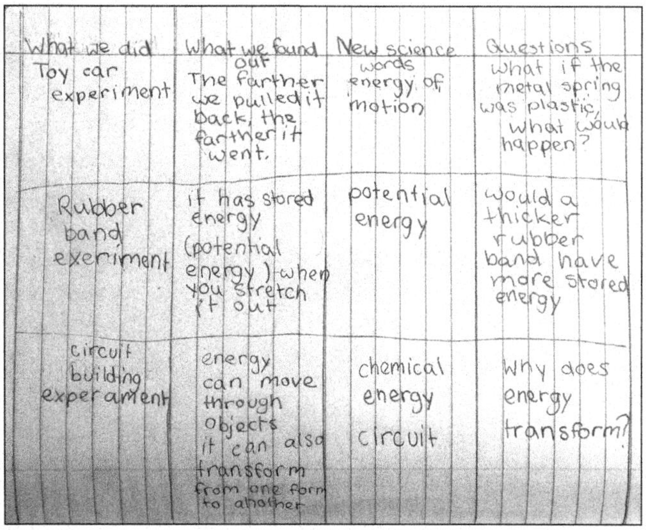

What we did	What we found out	New science words	Questions
Toy car experiment	The farther we pulled it back, the farther it went.	energy of motion	what if the metal spring was plastic, what would happen?
Rubber band experiment	it has stored energy (potential energy) when you stretch it out	potential energy	would a thicker rubber band have more stored energy
circuit building experament	energy can move through objects it can also transform from one form to another	chemical energy circuit	Why does energy transform?

Big Idea Pages

A similar strategy is the creation of big idea pages. I first learned about this strategy from my Twitter Professional Learning Network and was immediately intrigued by its possibilities for synthesis and explanation. As described by Jennifer Weibert, science coordinator for Fresno County School District in California, this notebook entry is a two-page spread with a central icon representing the topic and four real-life, phenomenon-based questions that direct student learning throughout a unit of study (Fresno County Superintendent of Schools n.d.). As the unit progresses, students collect evidence to answer these guiding questions

Figure 6.29. A teacher-created big idea page.

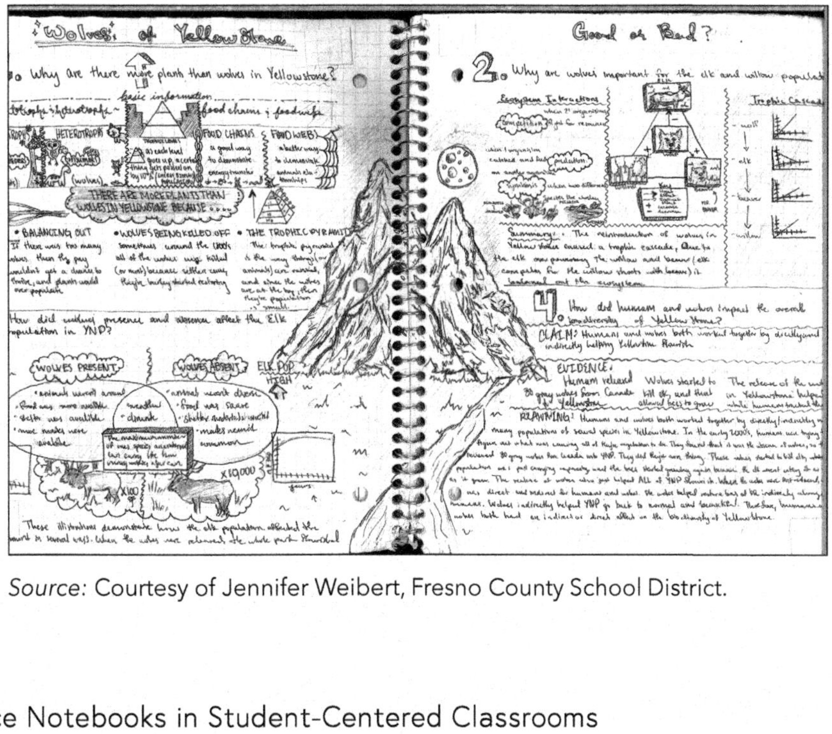

Source: Courtesy of Jennifer Weibert, Fresno County School District.

and record it in the appropriate space. Figure 6.29 (p. 77) shows an example of a teacher-created big idea page on the Yellowstone ecosystem.

I have not yet tried out this strategy with an entire class, but I have used big idea pages with a few students for extension and enrichment. This is a helpful way for me to test out a notebook entry that I'd like to try with an entire class the following year. I get an authentic look at how students respond to a task, and the students have an opportunity to deepen their thinking and are enthusiastic about serving as "experts" the following year with their classmates. Figure 6.30 shows a big idea page that one of my fifth-grade students created throughout a unit on the flow of matter and energy through ecosystems.

Figure 6.30. A fifth-grade student's big idea page from a unit on ecosystems.

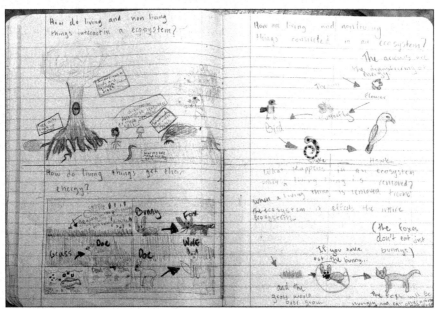

Reflective Paragraphs

A simple yet powerful way to support this science and engineering practice is to simply ask students to start by writing reflective paragraphs in their notebooks (Figure 6.31). In one example, a first-grade student's reflective paragraphs explain how the student determined that corn is a seed following the class's investigation, Why Is Our Corn Changing? (6.31a). In another, a third-grade student uses ideas about gravity to explain the movement of a furniture slider when her class conducted the Classroom Curling investigation (6.31b; Fries-Gaither and Shiverdecker 2012). And finally, a fifth-grade student draws on several pieces of evidence from in-class investigations and personal experience to explain the phenomenon of day and night (6.31c).

Figure 6.31. Examples of students' reflective paragraphs.

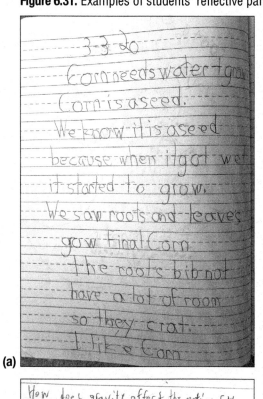

(a)

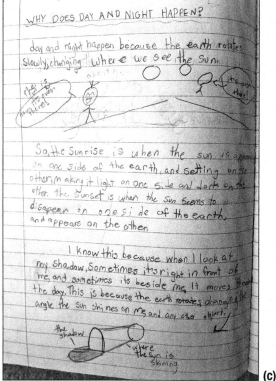

(c)

(b)

A first-grade student's explanation of their discovery that corn is a seed (a); a third-grade student's explanation of the motion of an object (b); and a fifth-grade student's explanation of the phenomenon of day and night (c).

Notebook Entries for Critiquing

Although they are important, students' own interpretations of data, models, and explanations are not enough. To fully develop a robust understanding of phenomena, students must compare their thinking with that of their peers and with the established body of scientific knowledge. Argumentation is one science practice that falls into this category, and claim and evidence T-charts and formal written arguments in notebooks support this practice.

A critical yet sometimes underemphasized science and engineering practice is the ability to locate, evaluate, and comprehend information and to clearly and concisely share ideas and information with others. Scientific advancements are only partially effective if they cannot be shared with the general public, and it is imperative that new discoveries and ideas be shared in a way that

Chapter 6

is comprehensible to those without scientific degrees. In terms of school science, specifically elementary school science, this means that students need to be able to read and comprehend science texts in all forms (including diagrams, infographics, and videos) and express their findings clearly and appropriately for a particular audience.

The cognitive demands of this practice present an exciting opportunity for elementary teachers to fully integrate science and other disciplines, including literacy, math, and social studies. Using an interdisciplinary approach to making sense of phenomena or solving a problem best reflects the way that science and engineering are conducted in the professional world and helps fully realize the potential of this practice. A few types of notebook entries for critiquing that align with this practice are sketchnotes, T-charts, research notes, and rough drafts and storyboards.

Claim and Evidence T-Charts

While very young students certainly make claims and support their ideas with evidence, I introduce formal documentation of claims and evidence at the beginning of second grade. A simple T-chart has proven to be an effective organizational tool for students in linking their claims to evidence. I initially provide graphic organizers, but students quickly become comfortable drawing their own T-charts. Figure 6.32 shows a student's claim and evidence following an investigation of whether air is a type of matter. Note the sketches that accompany the T-chart and give additional supporting evidence, as well as documentation of the experimental procedure.

Students become more sophisticated over time in their use of a T-chart as an organizational tool, as shown in Figure 6.33. A major difference, however, is that students begin to include more quantitative data that can be used as evidence. I often use a notebook workshop model (see Chapter 5)

Figure 6.32. A student's claim and evidence T-chart to answer the question "Is air matter?"

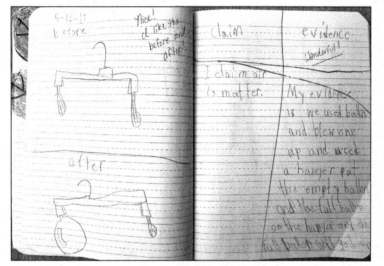

Figure 6.33. A fourth-grade student's claim and evidence T-chart shows the increasing sophistication in evidence.

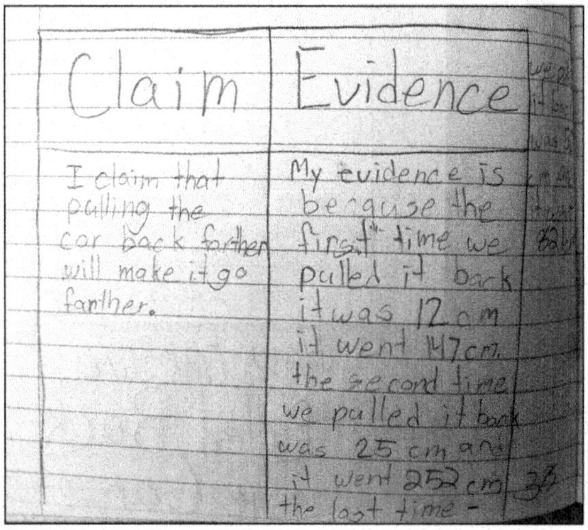

to support students as they learn to cite specific data as evidence for their claims. Students also frequently discuss the reasoning behind their claims, an important step toward writing a formal argument.

Formal Written Arguments

I introduce the concept of a formal written argument in fourth grade, after students have become proficient in supporting their claims with strong, data-driven evidence. During fourth and fifth grades, students gradually gain independence in constructing an argument using the claim-evidence-reasoning (CER) framework. I often begin by providing the framework for an argument that we have created in a scientist meeting, leaving blanks for students to fill in relevant evidence from their investigation and scaffolding the beginning of a reasoning section (Figure 6.34). We usually color-code the parts of the argument to serve as a model for future work. The next step toward independence typically involves composing and color-coding the parts. Eventually, students are able to compose their own arguments with less support and sometimes even begin to include a rebuttal section in addition to their claim, evidence, and reasoning (Figure 6.35). In my experience, students continue developing proficiency with argumentation well past the elementary years, particularly in terms of composing a strong reasoning section.

Sketchnotes

Also known as visual note-taking, sketchnoting is the practice of recording and responding to information in a combination of words and visual elements, including fonts, colors, symbols, and sketches. Educator and author Tanny McGregor (2018) explains that this practice is more than just a fun variation on note-taking: It makes the act of note-taking thinking-intensive. Choosing to use a combination of words and design elements is a record of a student's understanding of a text, video, or speaker. McGregor asserts (and I can attest from personal

Figure 6.34. A fill-in-the-blank, color-coded CER is an important scaffold for students gaining proficiency with argumentation.

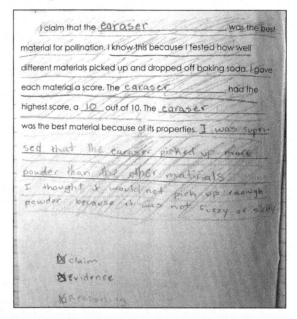

Figure 6.35. A fifth-grade student's argument shows increasing independence with argumentation.

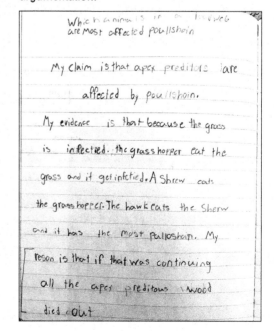

Figure 6.36. A fifth-grade student's sketchnote about the role decomposers play in an ecosystem.

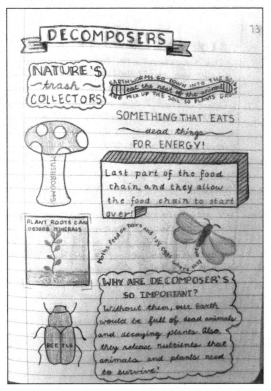

Figure 6.37. A fourth-grade student's sketchnote about the various forms of energy.

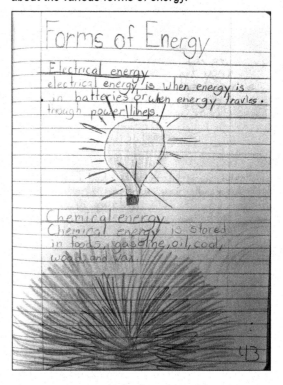

experience) that the practice allows for voice and choice, enhances student focus, and strengthens the memory.

Science notebooks are a perfect vehicle for incorporating sketchnotes. Figure 6.36 shows a sketchnote created by a fifth-grade student learning about the role of decomposers within an ecosystem. For this assignment, I gave students several sources of information about decomposers—a video clip, an informational text, and a poem—and invited them to synthesize what they learned into a sketchnote. As the sample work demonstrates, students threw themselves into this exercise with gusto, taking great pride in their sketchnotes and using a more thoughtful approach than if they had simply been asked to take written notes from the texts.

This was not my students' first attempt at creating sketchnotes; they had practiced the technique the year before with a single informational text on energy sources. Figure 6.37 includes an example of a beginning sketchnote, and in comparison with the previous example, it shows how students' work becomes more creative and sophisticated with familiarity and practice. As with any instructional strategy, modeling and practice help students feel more comfortable and take ownership of their work.

Sketchnotes can be used with a variety of informational sources—passages from textbooks, nonfiction articles, infographics, and even videos and live presentations. McGregor's 2018 book, *Ink and Ideas*, is a creative and practical guide if you wish to learn more and implement the strategy in your own classroom.

Other T-Charts

Besides the claim and evidence T-charts, T-charts can be used as graphic organizers for recording and responding to text. While a T-chart in and of itself is not a specific strategy, the intentional and careful use of prompts for each column can invite student reflection and response. Figure 6.38 shows two examples of T-charts used by my students as they read and reflected on texts. Note the pairing of the prompts in each example: What I Already Knew/New Information (6.38a) and What I Learned From Reading/How It Connects to Our Investigation (6.38b). In both cases, students were asked to do more than simply copy facts from the text: They were challenged to approach the text with a reflective, metacognitive stance. As with sketchnotes, T-charts can be used with all types of informational sources, and the prompts can be customized to fit the purpose of the activity.

Figure 6.38. T-charts used to promote reflection and critical thinking about informational text.

(a) (b)

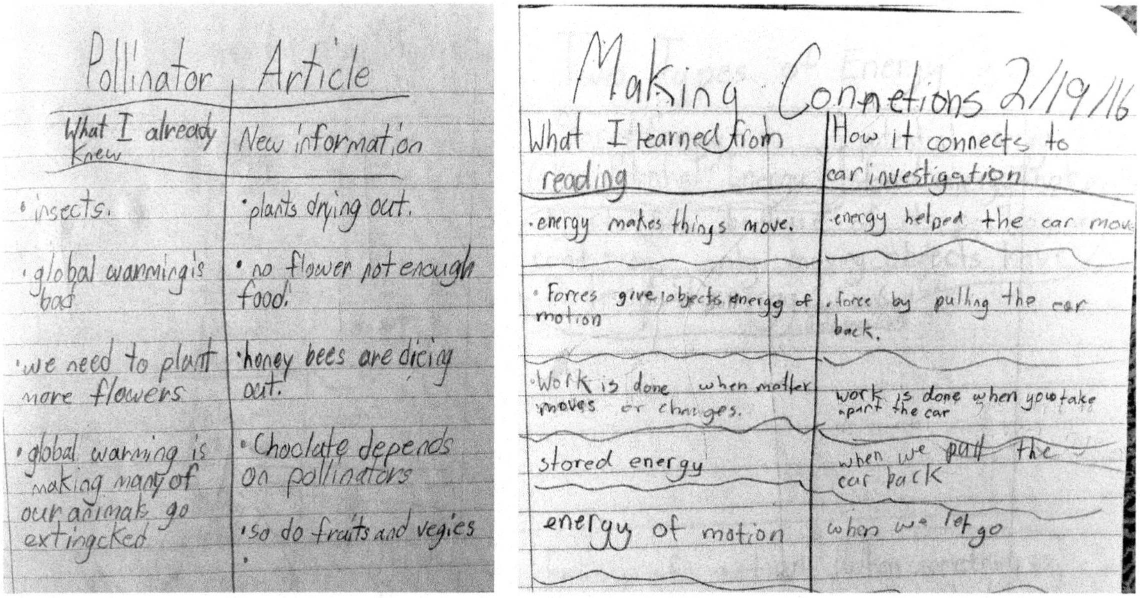

A fourth-grader's T-chart in response to an article about the global decline in pollinators (a); and a fourth-grader's T-chart showing connections between a hands-on investigation and an informational text about energy (b).

Figure 6.39. Examples of note-taking entries in students' notebooks.

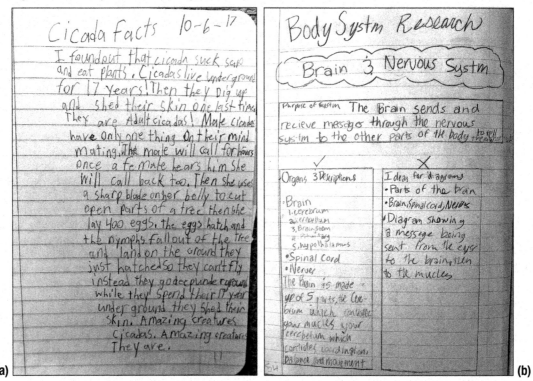

(a)

(b)

A third-grade student's entry noting important information from books about cicadas (a); and a fifth-grade student's entry about body systems for a formal research project (b).

Research Notes

Science notebooks can also be used for traditional note-taking and research. Figure 6.39 shows examples of different applications: an entry written by a third grader when I invited students to write down important information as they read books about cicadas as part of a life cycle investigation (6.39a) and a fifth-grade student's entry about a formal research project on body systems (6.39b). As with other types of entries, I try to balance student choice with scaffolding when specific outcomes are needed. In the third-grade case, these notes were simply the basis for a class discussion and question formulation session, so I was able to be open-ended in my directions. In the fifth-grade example, the result of this research was a cross-curricular essay for science and language arts classes. Note that the *NGSS* has moved this science content into middle school (grades 6–8). I thus chose to provide more scaffolding in this instance, creating a list of topics for students to research and guiding them in drawing their own graphic organizers in their notebooks.

I've found that some students gravitate to this practice of note-taking and ask to use their notebook during read-alouds, videos, and other times when I haven't asked students to do any writing. I'm more than happy to oblige, as I know that they are discovering their own preferences with regard to learning.

Rough Drafts and Storyboards

To fully realize this science and engineering practice, consuming information from others is not enough. Instead, it is equally important (if not more so) that students are able to produce explanations and create products to share their new understandings. Science notebooks provide a safe space for students to craft an explanation or plan an infographic or a multimedia presentation, receive feedback, and improve their work—all before actually sharing the work with the intended (and whenever possible, authentic) audience.

For written work (paragraphs, essays, poems), I ask students to write their rough drafts in their notebooks (Figure 6.40). We often follow a traditional writing process in which students share their drafts with peers, revise, receive feedback from me, and finally "publish" their work in a final draft that appears outside of their notebook. This final product might be written neatly on notebook paper or typed and printed, depending on the age of the students and the specific details of the assignment. Regardless of the format, I strive to make a clear distinction between the students' notebooks as a thinking tool and the more polished work that is ready to be shared with the public.

In other situations, students are tasked with creating different types of products, such as infographics, videos, or multimedia presentations. Notebooks become a place to brainstorm, plan, and storyboard before actually working with the digital tools needed for production. Figure 6.41 shows a student's rough draft of the information the student wanted to share in a Flipgrid video.

Figure 6.40. A student's rough draft of an informational text about forces.

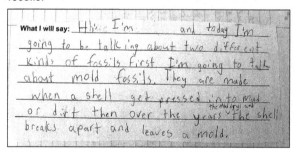

Figure 6.41. A student's script for a Flipgrid video on fossils.

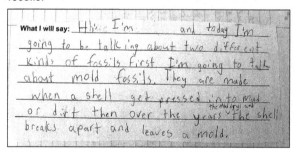

As with student writing, I feel it is essential to preserve the notebook as a safe space for the iterative work involved in the creation of the final product.

Crosscutting Concepts

To truly reflect the three-dimensional nature of science instruction as called for by the *NGSS*, science notebook entries should also provide opportunities for students to engage with the seven crosscutting concepts: patterns; cause and effect; scale, proportion, and quantity; systems and system models; energy and matter; structure and function; and stability and change. These concepts serve as a mental framework or lens through which students can make sense of phenomena (see

Figure 6.42. A student's reflection on the patterns she has identified in maps of landforms, volcanoes, and earthquakes around the world.

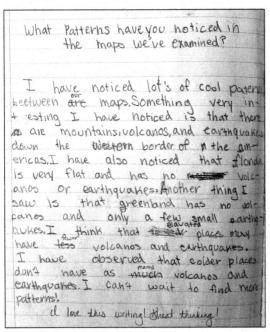

Figure 6.43. An exploded diagram of a friction car helps a fourth-grade student think about the crosscutting concept of systems and systems models.

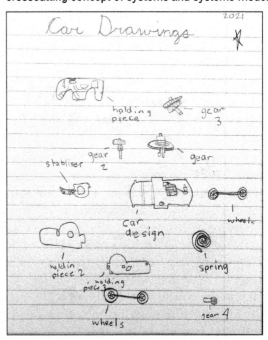

Chapter 1). This may happen explicitly through questions, writing prompts, and graphic organizers, or more subtly through the nature of the activity or the notebook entry itself. In this section, I share several examples of notebook entries and discuss how one or more crosscutting concepts are woven into the work.

In Figure 6.42, a student describes the patterns she observes in the location of landforms (such as mountains and volcanoes) and earthquakes. This work builds on the See/Think/Wonder thinking routine depicted in Figure 6.5 (p. 62). While many students start to identify patterns in their thinking routines on their own accord, others need the explicit language of the writing prompt to help them reach that conclusion.

An investigation that addresses multiple crosscutting concepts is a fourth-grade foray into energy using a pullback car. Students work in collaborative groups to plan and conduct an experiment to answer the question "Will the car go farther if we pull it back farther?" After students conclude their experiment, they are invited to take apart their cars and determine how they work. This is a highly engaging part of the investigation, and as students excitedly discuss their discoveries with me and with each other, I hear them naturally use several CCCs: cause and effect, systems and system models, energy and matter flows, and structure and function. To help students think about these concepts more explicitly, I ask them to draw an exploded diagram of the parts of the car (Figure 6.43) and to answer several questions after discussing their thinking and reading an introductory text about stored energy and energy of motion:

- How do the parts of the car work together to make it move? (structure and function, systems and system models)
- What causes the car to travel farther when you pull it back farther? (cause and effect)
- How does the car's energy change from stored energy to energy of motion? (energy flows)

Student responses to these prompts (Figure 6.44) show an emerging understanding of the cross-cutting concepts underlying this investigation.

Figure 6.44. Fourth-grade students' responses to teacher-posed questions as they begin to consider the role crosscutting concepts play in the movement of a pullback car.

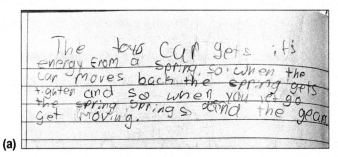

(a)

(b)

(c)

One student's response shows her understanding of structure and function as well as systems and system models (a); another shows an understanding of cause and effect in this investigation (b); and a third response shows a developing understanding of energy flows (c).

Chapter 6

Working explicitly with the CCCs in my instruction is admittedly an area of continued professional growth for me. I hope that with scaffolded opportunities to use the CCCs throughout the year, students will recognize which of many CCCs will help them make sense of a phenomenon and begin to use the CCCs independently.

The examples in this chapter demonstrate that science notebooks can successfully support a three-dimensional approach to instruction, as disciplinary core ideas, science and engineering practices, and crosscutting concepts are naturally interwoven throughout. Although I have described my typical approach to notebooking and shown a variety of student work, these samples show only a fraction of the wide range of student abilities in my science classroom. In the next chapter, I share strategies that have been helpful in meeting my students where they are and moving them forward at all grade levels.

References

Bragg, L. 1959. *A short history of science: Origins and results of the scientific revolution.* Chap. 15, The atom. Garden City, NY: Doubleday.

Fresno County Superintendent of Schools. n.d. Big ideas pages. *https://stem.fcoe.org/big-ideas-pages.*

Fries-Gaither, J., and T. Shiverdecker. 2012. *Inquiring scientists, inquiring readers: Using nonfiction to promote science literacy, grades 3–5.* Arlington, VA: NSTA Press.

McGregor, T. 2018. *Ink and ideas.* Portsmouth, NH: Heinemann.

McNeill, K., R. Katsh-Singer, and P. Pelletier. 2015. Assessing science practices: Moving your class along a continuum. *Science Scope* 39 (4): 21–28.

Museum of Science. 2020. The best of bugs: Designing hand pollinators. *https://eiestore.com/the-best-of-bugs-designing-hand-pollinators.html.*

National Governors Association Center for Best Practices and Council of Chief State School Officers (NGAC and CCSSO). 2010. *Common core state standards (mathematics).* Washington, DC: NGAC and CCSSO.

Next Generation Science Storylines Team. 2019. Why is our corn changing? v. 2.1. Next Generation science storylines. *www.nextgenstorylines.org/why-is-our-corn-changing.*

Phenomenal Science Team. 2017. Driving question boards. Phenomenal Science K–5 curriculum. *http://phenomscience.weebly.com/blog/drivingquestionboards.*

Project Zero. 2019. See/think/wonder. Harvard Graduate School of Education. *https://pz.harvard.edu/sites/default/files/See%20Think%20Wonder.pdf.*

CHAPTER 7

Supporting and Extending Students

The media need superheroes in science just as in every sphere of life, but there is really a continuous range of abilities with no clear dividing line.

—Stephen Hawking (quoted in Solomon 2004)

The beauty of student-centered science notebooks is that they provide a clear picture of students' understanding of science concepts and vocabulary—as well as their proficiency with science and engineering practices and associated literacy and math skills. Careful study of notebook entries thus allows you to meet your students where they are and move them forward through purposeful, scaffolded instruction. In this chapter, I highlight strategies for supporting students who need additional assistance and strategies for extending students who need greater challenge. You'll likely notice a common thread woven throughout these strategies—almost all are simply effective teaching practices for *all* students. Providing additional support might mean using the strategies for a longer period of time or more intensely, and providing greater challenge might mean removing some of the scaffolds earlier in the process.

Strategies for Supporting Students

Premade Elements

Teacher-created graphic organizers and other notebook elements support all students as they become familiar and comfortable with various thinking routines and notebook entries. For most students, I follow a gradual release of responsibility, moving from giving them a premade element to modeling how to create an organizer and eventually to having them create these elements independently. However, not all students progress at the same pace through this process. In some cases, students may need premade elements for a longer period of time or more direct support as they work to create their own entries.

Figure 7.1. Teacher-created sentence starters and sentence frames are reference entries added to students' notebooks at the beginning of the year.

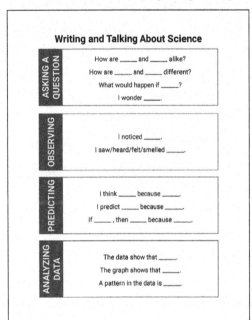
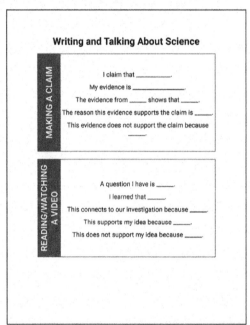

Sentence Starters and Sentence Frames

Over the years, I've come to believe that sentence starters (the first few words of a sentence) and sentence frames (a sentence with key components omitted) have an almost magical power in helping students move from experiencing writer's block to having something to say. It truly is amazing how just having a few words will unlock student thinking. I provide these tools for all students in a variety of ways: on anchor charts in the classroom, in reference entries at the beginning of students' notebooks (Figure 7.1; see also pp. 125–126 of the appendix), and in written and verbal directions for a particular assignment.

While sentence starters and frames are beneficial to all students, they can be used more explicitly as support for students for whom writing (and talking) about science content is a challenge. This may take a variety of forms, depending on the students' grade level, the activity in question, and the level of support required. In some cases, students may use teacher-created organizers with pretyped sentence starters and frames, as shown in Figure 7.2. In other cases, I may handwrite a sentence starter in a notebook to help the student get started or give a verbal suggestion during a one-on-one conversation. Since my ultimate goal is student independence in their notebook entries, I let students attempt an activity first and then follow up with support as needed whenever possible.

Checklists

Checklists are another tool that I use with all students, particularly for notebook entries involving observational drawing, graphing, or explanation (see Chapter 6 for several examples of checklists that help ensure student success in these types of entries). I frequently require students to use checklists that pertain to content or essential elements of a notebook entry. For other, more procedural tasks, like graphing, I start all students off with checklists and

Figure 7.2. Sentence starters provide support for an assignment.

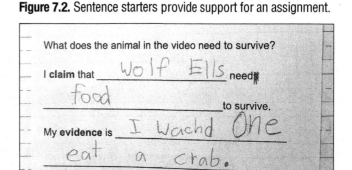

allow them to determine whether and when they are ready to engage in a task without one. In this way, students have choice in the amount of support they receive.

I also use a more informal type of checklist to support students who have challenges with executive functioning skills and completing multistep directions with regard to notebook entries. I keep laminated blank checklists easily accessible in my classroom so that I can quietly offer one to any student I notice having a difficult time beginning an assignment. Together, we write in the steps of the assignment with a dry-erase marker, and the student can check off the steps as they are completed (Figure 7.3). As with other strategies, the frequency of use depends on student needs. I always offer some students the use of a checklist, per their learning plans, but tend to use a wait-and-see approach with others. I introduce the checklists (along with

Figure 7.3. Laminated checklists can serve as tools for specific assignments and can be erased when completed.

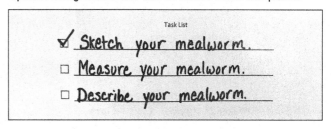

other, similar tools) to students at the beginning of the year and let them know that they are always welcome to use these tools during class—again allowing for choice concerning support.

Individual Conversations and Notebook Conferences

A great deal of my support for students happens during informal, one-on-one conversations as they work in their notebooks. Over-the-shoulder feedback gives me an opportunity to ask questions about their work, share my observations, and support or extend as needed. Sometimes, though, a student requires more assistance than I can give them in a 30-second conversation. In these cases, I set up a notebook conference, where I can meet with a student for an extended conversation and provide more guided use of some of the strategies described here. These conferences can happen in the classroom while other students have moved on to a different activity or after class if more privacy is warranted.

Chapter 7

Reference Elements and Exemplars

In addition to sentence stems and frames, student notebooks also contain other reference elements, including a self-created master vocabulary list (alphabet boxes) and student-generated definitions for key vocabulary terms, or a glossary for grades 3–5 (see Chapter 5 for more details on these items). Students are encouraged to refer to these elements at any time to enhance their work or remember how to spell a word correctly.

I also have students create exemplars in their notebooks for future reference. Figure 7.4 shows a bar graph made by a fifth-grade student at the start of the year with a non–standards related activity (quantities of Skittles by color). I assisted individual students as they completed their graphs and reviewed their work upon completion to be sure that it was correct and included all essential elements. Other exemplars that students have created as notebook entries include whiteboard models and arguments color-coded by claim, evidence, and reasoning. Throughout the remainder of the year, students can refer to these exemplars, either on their own or in response to my feedback.

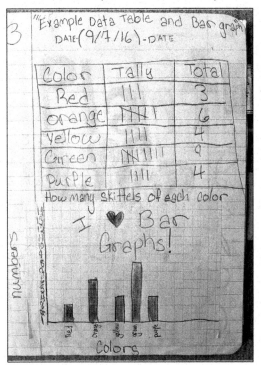

Figure 7.4. Student-created bar graph entry that serves as an exemplar for the rest of the year.

Voice Recordings and Teacher-Scribed Entries

Throughout my years of teaching, I've worked with quite a few students who have terrific ideas in their heads but struggle to get them down on paper, and you likely have as well. Whatever the underlying cause of this challenge, it makes completing notebook entries quite frustrating for students. While switching to a digital notebook format can help in some cases, it doesn't always address the issue and may not be possible if technology is limited. So how can you support students with pencil-and-paper tasks? I've found success with two strategies: voice recordings and teacher-scribed entries.

When I can tell students are struggling with a written notebook entry, I have a quick conversation with them. If it seems that they have the ideas but are having difficulty transferring their thinking to paper, I may give them a recording device and send them into the hall to record themselves answering the prompt out loud. They can then play back their recordings and transcribe their responses at their own pace, as the video or audio files have saved their in-the-moment thinking. Audio recording apps, the video recorder available through most mobile devices, or even Flipgrid are all possibilities for doing this. This modification does add time to the assignment, so I also build in a flexible deadline or set a time for these students to come in and work outside of class. Audio and video files can also be saved to the cloud or emailed to students, giving them the ability to work at home. Note that in some instances, it is not appropriate or reasonable to expect a student to do this, and an audio recording or video will suffice without a written transcription. However, this is one way I meet students where they are while also giving them the opportunity for continued practice with writing.

Another strategy that I reserve for students with significant challenges is scribing entries to some degree. As with other strategies, this can take many different forms. I might draw or assist with an organizer or framework for students to fill in, or I might write a sentence frame, omitting key words and asking students to supply them. In some situations, writing the first sentence of an entry can provide the needed push for a student to begin working independently. Occasionally, I may scribe an entire entry, adding a note so that the students, their family members, and I will all remember that this modification was used. Figure 7.5 shows an example of an entry by a student for whom writing was a slow and laborious process. She was running out of class time and I didn't want her to lose her thinking, so I scribed the last few sentences of her entry.

Figure 7.5. A fifth-grade student's notebook entry showing a section that was scribed by me as a form of instructional scaffolding.

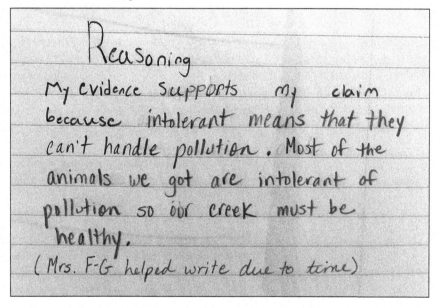

Scribing is generally the last strategy I try, but for some students, it means the difference between a response and a blank page. Since my primary goal is to facilitate students' sensemaking of phenomena and understanding of science concepts, I am willing to do what is needed to overcome barriers that interfere with student success. I then gradually move the student toward independent work over the course of the year, removing small supports with each success.

Partner Support

A final yet sometimes underrated support strategy is to allow students to work with more capable partners on a task. Sometimes my best attempts at explanation simply do not resonate with a student, and only a peer can explain a question or a concept in a way that makes sense to that student. This type of work requires monitoring to be sure that both students are contributing to the task, and it is generally not effective for notebook entries that will be used for assessment. However, my students' notebooks are primarily a source of formative assessment data (see Chapter 8), which means that partner and group work is generally OK.

Strategies for Extending Students

Although many students in our classrooms need additional support for academic tasks, a not-so-insignificant number are ready for greater challenge. Such students are often continually asked to help their peers, but they deserve the opportunity to grow and develop in their scientific thinking and practices. Meet students where they are and move them forward.

I find that, generally speaking, science notebooks promote authentic self-differentiation, in that students engage with the tasks at their own level of understanding and proficiency. This means that students naturally extend themselves through greater detail, more sophisticated reasoning, and deeper understanding of concepts. Open-ended prompts that allow for voice and choice maximize the chances of this happening. Nevertheless, my goal as an educator is to extend these students even further. While I don't have a lengthy list of specific strategies to share, several have proven successful: individual conversations, partner work, and, when appropriate, independent projects.

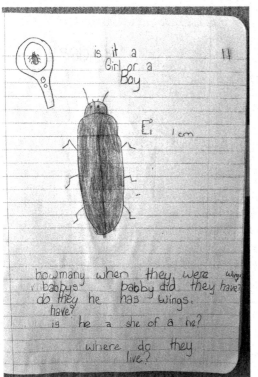

Figure 7.6. A third-grade student's observational drawing of a darkling beetle, which was modified following my one-on-one conversation with her.

Individual Conversations

Over-the-shoulder feedback is one of my most frequently used strategies for both supporting and extending students. As I watch students at work, I can provide real-time comments and questions to help them dig deeper. Figure 7.6 shows a third-grade student's observational drawing of a darkling beetle, and my initial assessment of her work was that she had mastered the basics of documenting her observations. I stopped by her desk and had a short conversation about looking more closely at the beetle's legs to determine whether they were straight lines, as she had originally drawn, or jointed. We also talked about ways to document that she had drawn the beetle larger than life-size; in the end, she opted to add a centimeter scale for reference. In other cases, I've challenged students to increase their precision in description and measurement.

Other times, I may use one-on-one conversation to press students to think more deeply about a concept and elaborate in their writing. The crosscutting concepts serve as an excellent vehicle for student extension, because even if the class has focused on a specific CCC, several others typically align with the topic at hand. Inviting students to think about other CCCs that could be used to explain the phenomenon they've studied can be a rich opportunity to delve deeper.

Finally, I use these individual conversations to encourage students to think more deeply about experimental procedures, data, and possible sources of error. For some, I may introduce a mathematical computation (such as determining the mean) that they are ready to try before their peers. In other cases, I ask them to think about how an investigation might be changed to reduce sources of error or better control variables.

Partner Work

While conventional teaching wisdom supports grouping high-ability students with peers who could use assistance, I regularly look for opportunities to pair these students together. I've found that they not only enjoy the chance to be a learner instead of a helper but also feed off the ideas and creativity of their similarly matched partner. The end result is typically more thoughtful and meaningful than either student would have produced individually.

Independent Projects

The strategies for support and extension that I've discussed up until now can be applied in most situations. This strategy, however, is one that I resort to only in certain cases. There are times when I've found that individual students have already studied a topic we are about to investigate, and through conversation, I can tell that they won't gain much from participating with the class as a whole. Typically, these students have moved to our school from another area, resulting in a curricular mismatch. When this occurs, I feel that the best option is to provide an opportunity for these students to work on an independent project. I meet with such students outside of class to determine a project topic, plan of investigation, and final product and let them use their notebooks to document their learning. Their independent project replaces the work they would otherwise do, and I find ways to assess similar practices and skills in their work as with their classmates.

Student-centered science notebooks allow students a chance to work at their own comfort level in terms of both content and process and thus provide a natural opportunity for differentiation. It is essential to remember that all student work has value and that there is always something to commend and praise, even if there is much room for growth. It is also important to reflect on the types of strategies used to extend students: Are students going deeper into their work as scientists or just doing *more* work? The former is what will keep those students learning and growing, while the latter can be demotivating. Ultimately, careful, purposeful assessment of student work in science notebooks is needed to leverage these opportunities and strategies. In the next chapter, I discuss how I use science notebooks for both formative and summative assessments and how I have students share their work with a broader audience.

Reference

Solomon, D. 2004. "The Way We Live Now: 12-12-04: Questions for Stephen Hawking; The Science of Second-Guessing." *New York Times*, December 12.

CHAPTER 8

Assessment and Sharing of Science Notebooks

In questions of science, the authority of a thousand is not worth the humble reasoning of a single individual.

—Galileo Galilei (quoted in Misner, Thorne, and Wheeler 1973)

As an ongoing record of students' development and growth, science notebooks are a virtual treasure trove of formative and summative assessment material. You can quickly scan students' work to confirm their mastery of a lesson's objective, delve more deeply to determine the instruction they need over the days and weeks to come, or analyze a single entry to assess their understanding of and proficiency with science practices and skills. If anything, the real challenge is not in having enough material to assess but in keeping up with the volume of student work. While I certainly cannot claim to have completely solved this problem, my years of teaching have forced me to think creatively when it comes to managing notebooks. I hope some of these suggestions prove helpful for you. Though not directly related to assessment, I also describe my approach to solving another challenge posed by the use of notebooks: how to share student work and growth on a regular basis with parents and guardians.

First, let me give some background as context for my approach. I teach at a school in which the elementary grades are standards-based. We distribute formal written reports twice a year in which we document and describe student progress on subject-based indicators using a four-point descriptive scale: beginning, developing, secure, and exceeds expectations (Figure 8.1). In addition to the indicators, we write a narrative report with other information not covered by these indicators. Math and literacy indicators are specific, discrete, and numerous, whereas those

Figure 8.1. Four-point assessment scale used at my school.

> **Beginning:** Beginning understanding of grade-level expectations; consistently requires assistance.
>
> **Developing:** Developing understanding of grade-level expectations; requires some assistance.
>
> **Secure:** Secure understanding of grade-level expectations; works independently at grade level.
>
> **Exceeds expectations:** Extending understanding and application; advances beyond grade-level expectations.

used in other classes (including science) are broader statements of skills and are fewer in number. These formal progress reports are supplemented with parent-teacher conferences and all the typical informal ways in which teachers communicate with parents.

This assessment landscape has challenged and shaped my thinking over the years—to the point where I would be hard-pressed to return to a numerical system of grading. While I don't mean to imply that science notebooks are successful only within the framework of a standards-based system, I feel that understanding the particulars of my situation provides a lens through which to read this chapter.

Whether to Grade Notebooks

The question of whether to grade notebooks is not truly applicable to me, as a teacher in a standards-based school, but it is one that I am often asked and have strong opinions about. In short, I do not recommend attaching a numerical grade to a notebook as a whole, even if you teach in a school that assigns numerical grades. Attempts to do this result in a checklist of required elements and formatting conventions, including criteria such as all pages numbered, entries properly headed, and the subjective "neat and organized." These have little, if anything, to do with students' development of conceptual knowledge and abilities to transfer skills and ideas and result in a compliance mindset toward the notebook instead of a creative one. A numeric grade doesn't capture the developmental understandings that students gain. Students' growth is more important than a standard achievement score.

At first, this may seem contradictory to my efforts to teach my own students to organize their notebooks, and it is a point that deserves further explanation. Although I do ask students to include organizational elements in their notebooks and provide scaffolded approaches to help them learn to organize their work independently, I don't see this as an area to grade. If you were to browse through my students' notebooks, you would see a variety of levels of organization, even with my instructions and modeling. Not every entry is properly titled, sometimes entries end up out of order, and work is often messier than I would like. Does that mean students aren't engaged in class or aren't learning? Not at all. That is the reality of teaching elementary students who develop at their own rates. I set a goal for their work with my organizational schema, but in the end, I meet students where they are and try to move them forward. This means giving verbal and written feedback on organization as well as content and creating a classroom environment in which students can make mistakes and learn from them. Attaching a numerical grade to something like organization would simply penalize students who were continuing to develop in this area.

Additionally, I believe that teachers should avoid grading individual notebook entries for content as much as possible. My philosophy is that notebooks are a safe space for student growth and learning, much as practice is a place for athletes to develop skill in a sport. Notebooks help students develop and refine skills so that they perform to their best abilities in an assessment situation. Adding an evaluative component in the form of a grade violates that safe space for practice. I add written comments to student work to help them improve future entries and help me plan my instruction, but over the years, I've discovered that I prefer to keep my summative assessments outside of the notebook as much as possible. In the following sections, I discuss ways that I use notebooks as sources of formative assessment and the basis for summative assessment.

Formative Assessment

My students' notebook entries are valuable opportunities for formative assessment. I do this in a number of ways: in-class checks, after-class review, and embedded exit tickets. During independent or collaborative work time, I circulate through my classroom, talking with students and peeking over their shoulders at their notebook entries. These on-the-fly observations often let me know if I should stop the class to clarify directions or a concept or if a student or group of students may need additional support. As my classroom is a hive of activity during these times, I walk around with a class list, checking off student names as I peek at their notebooks. This ensures that I see all students' work in progress and allows me to jot down notes for later reference.

Real-time formative assessment is terrific, but it has its limits, especially with longer, more writing-focused entries. In these cases, I prefer to review student writing after class has ended. This can give me insight into students' growing understanding of content and use of vocabulary. However, as a specialist responsible for teaching multiple grades, it simply isn't possible to review every notebook entry after every class. I pair my observations of students at work and their contributions to our class discussions as a first line of formative assessment, choosing to review specific students' work based on my in-class evaluation. If a large percentage of the class had difficulty with an assignment, or if we worked on a particularly tough concept, I review all entries from that day; otherwise, I have to prioritize my time and attention.

Another way I use my notebooks for formative assessment is by embedding exit ticket–type tasks in notebook entries, in which I ask students to answer a question or write a sentence or two in response to the lesson's content. These prompts might be content-specific, or they may be general, like the following:

- "Write down the most important thing you learned from our reading today."
- "What one question do you still have after today's class?"
- "Summarize what you learned in one sentence."
- "What goal do you have for next class?" (in the case of independent projects)

Students complete their exit tickets on the same page they were using in class. Depending on the amount of time remaining in the period, I may walk around and check student work, responding on the spot. If time is running short, I ask students to mark that place with the bookmark so I can review their work later on, then I try to comment in writing on each student's work by the next class.

In some instances, it is more efficient to supply each student with a paper copy of an exit ticket. I often ask students to glue these into their notebooks once the tickets have been assessed and returned. Figure 8.2 (p. 100) shows an example of an exit ticket assessing students' proficiency in classifying questions as observation, research, or testable following a QFT-focused lesson. Student performance was not part of a formal assessment but helped me determine what instructional

Figure 8.2. Students add completed exit tickets to their notebooks as a permanent record of their growth throughout the year.

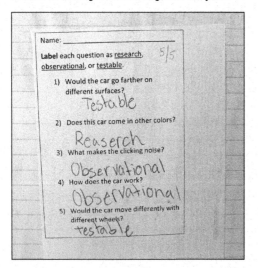

activities to plan for the following lesson. Adding these exit tickets to notebooks really documents the progression of student understanding throughout a unit or the entire year.

Summative Assessment

Although summative and formative assessments serve a different purpose, the notebook can often be used to accomplish both, depending on the task. Several times a year, I report students' progress to parents on a set of indicators, many of which are practice-focused. For these indicators, notebook entries are a natural assessment source. With regard to an argumentation indicator, for example, I use selected claim and evidence statements in students' notebooks as data points to determine where students fall on our descriptive four-point scale (beginning, developing, secure, exceeds expectations). To be clear, not every claim and evidence statement ends up serving as an assessment point, since much of what students do in their notebooks is practice. The idea of the notebook as a practice space is something that I establish early on in the school year with students, so in the rare case when I plan to use a notebook entry for a summative assessment, I tell them in advance so they can approach the assignment accordingly. And the feedback that I provide to students is almost identical whether I'm formally assessing their work or not.

When it comes to assessing student understanding, I prefer not to rely on notebook entries but instead use products outside of the notebook. Sometimes students compose rough drafts of explanations in their notebooks, then "publish" their final drafts outside of their notebooks as a summative assessment. Other times, notebook entries become a source of information for a traditional quiz or test. By keeping the assessment separate from the notebook as much as possible, I hope to preserve the notion of the notebook as a safe space for thinking through ideas, giving students freedom to explore and experiment without the fear often attached to assessment and evaluation.

One area of three-dimensional instruction and assessment in which I continue to develop and grow as an educator is in using the crosscutting concepts in assessment tasks, whether in or outside of student notebooks. While I use the CCCs to create some prompts for written work and for questions on tests, I know that students need to do more with them.

Assessment and Feedback Tools

Rubrics and checklists of various forms are incredibly useful in assessing notebook entries for both practice and content. I use some of these tools for my own assessment and record-keeping, and I share others directly with students. Four-point descriptive rubrics, like the one shown in Table 8.1, help me consistently assess student growth on my course's indicators (practice and skill statements). As the language and concept used in this rubric is sophisticated, I don't share this

Table 8.1. Four-point descriptive rubric for assessing students' claims and evidence.

	Exceeds expectations	**Secure**	**Developing**	**Beginning**
Claim	Generalizes to broader concept	Uses comparative language or otherwise draws a conclusion	Restates finding from experiment	Is incorrect for content
Evidence	Includes multiple data points or uses mathematical calculations to analyze data	Includes sufficient data (typically two points for comparison) or qualitative data as appropriate	Includes only one data point or uses comparative language without data	Explains why or says, "I did an experiment"

rubric with students. Instead, I use written and oral feedback to help guide students to the next level of proficiency. This type of rubric can also be helpful when communicating student progress with parents and guardians.

Other assessment and feedback tools are more student-friendly. One of my favorites, used for observational drawings and descriptions, is a simple table printed on a sticky note, as shown in Figure 8.3. The criteria in the table are ones I have used to guide students' observations: that they are clear, complete, and accurate; use scientific terminology ("science words"); and use objective language ("fact words"). I write an *X* in one of the boxes for each criterion to show students whether they are above, at, or below the grade-level expectation for that criterion. (Students are already familiar with these criteria and can refer to exemplars on anchor charts in our classroom as they work.) While I typically use these for feedback and not as an actual assessment, they could be easily modified for that purpose. (Templates and printing instructions can be found in the appendix.)

Another tool I share with students for feedback and assessment is a checklist. Similar to the checklists I use to scaffold student work, these list the criteria to be considered "secure" on a given assignment. Figure 8.4 (p. 102) shows one such checklist that I use for student graphs. Having criteria laid out in this manner simplifies my assessment process and provides clear guidance for students and their families in terms of needed improvements.

Figure 8.3. Sticky notes are a convenient way to share assessment and feedback with students.

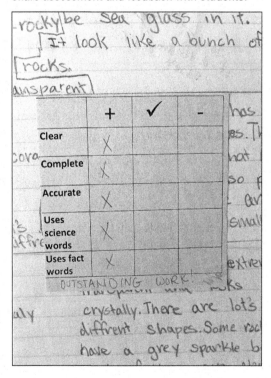

Figure 8.4. Checklist detailing requirements for a student to be "secure" in creating a bar graph.

Bar Graph Checklist

☐ The graph has a title.
☐ The intersection is labeled with 0.
☐ The x-axis is labeled.
☐ The y-axis is labeled.
☐ Numbers on the y-axis are labeled on the lines, not the spaces.
☐ There are equal intervals on the axes.
☐ There is a space between the y-axis and the first set of data.
☐ There is equal spacing between each set of data.
☐ If there is a zero amount, it is represented with a line.
☐ Double and triple bars are labeled with a key.

Finally, I have found single-point rubrics to be incredibly useful in assessing student work that is more content-focused. I first learned about single-point rubrics in a blog post on the Cult of Pedagogy website (Gonzalez 2015). While the author is quick to point out that she did not invent this type of rubric, she makes a convincing argument for their use in the classroom. I began implementing these rubrics with my students and found them to be easier to create and explain to students, and they are aligned with my school's assessment philosophy. Figure 8.5 is a single-point rubric for a fifth-grade summative assessment in which students constructed an explanation of observed relationships in a food web. Students first drafted this assignment in their notebooks, revised their writing, and then published their final drafts outside of the notebooks.

Figure 8.5. Single-point rubric for a summative assessment explaining the observed relationships in a food web.

Exceeds expectations Evidence of exceeding standard	Secure at grade level *Standard for this assignment: Constructs an explanation of observed relationships in the prairie food web*	Developing Areas that need work
	Explains that a food web models the flow of matter throughout the ecosystem	
	Explains that food of almost any kind of animal can be traced back to plants	
	Explains that plants get their energy from the Sun through *photosynthesis*	
	Explains the relationship between *herbivores* and *carnivores*	
	Provides a specific example from the prairie food web	

Sharing Notebooks With Parents and Guardians

One major challenge with a heavy reliance on notebooks is that students' work stays at school most of the time. I store science notebooks in my classroom and do not typically assign homework, per school policy, so students have little reason to take them home. The result is that unless a student shares detailed accounts of their school days, parents and guardians are likely to be unaware of the terrific thinking and learning happening in my classroom. I have found success in planning for the regular sharing of notebooks.

At the end of each unit, I invite students to review and reflect on their notebook entries, selecting a few entries that they are particularly proud of, one entry that reflects something new that they learned, and an area in which they would like to improve. Students write about their selections on bookmarks (Figure 8.6) and then take their notebooks and bookmarks home with the assignment to share their work with their families. Parents or guardians add comments to the bookmarks and sign them, and students return the materials to school. I have found that students take great pride in sharing their work, and the adults enjoy the chance to see what happens in our classroom on a daily basis. (A blank copy of this bookmark is included in the appendix on. p. 130 for your use.)

Parent-teacher conferences are a perfect opportunity to share science notebooks. Having concrete work examples helps focus the conversation, and they serve as clear indicators of what students are doing well and where they could improve. Sometimes, I invite parents to leave a comment (written on a sticky note) about a notebook entry they particularly enjoyed viewing. It's always a pleasant surprise for students to open their notebooks the next week and find a note about something they've done well.

Figure 8.6. Bookmark for reflection on and sharing of notebooks with parents and guardians.

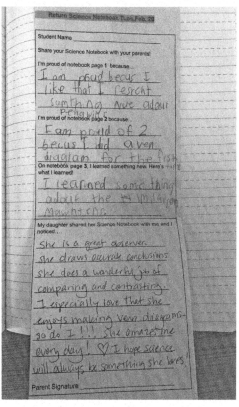

Creating a digital portfolio is another possibility for sharing notebook entries and one I hope to explore in the future. In this approach, students select work to photograph and place on a digital platform, such as Google Slides. This work could demonstrate mastery of a certain standard or simply be work the student is proud of. Students can add to portfolios throughout the course of the year and share them with parents at student-led conferences.

♦ ♦ ♦

Chapter 8

With all the time, energy, and effort that students put into their science notebooks, it's no surprise that they are terrific forms of assessment. This is particularly true in a standards-based classroom, where students are working to demonstrate mastery of standards and indicators. Although notebooks afford assessment opportunities in classrooms with traditional grading, you should reserve grading for content and not compliance with notebook conventions. Notebooks can also play a large role in shaping your instructional decisions. In the final chapter of this book, I share how selected student entries shaped my decisions as an elementary science specialist.

References

Gonzalez, J. 2015. Meet the single-point rubric. Cult of Pedagogy. *www.cultofpedagogy.com/single-point-rubric/*.

Misner, C. W., K. S. Thorne, and J. A. Wheeler. 1973. *Gravitation.* San Francisco: W. H. Freeman.

CHAPTER 9

Science Notebooks in Practice

I like to learn. That's an art and a science.

—Katherine Johnson (quoted in Clarke 2020)

In the preceding chapters, I've shared my philosophy on the place of science notebooks in a student-centered, three-dimensional classroom as well as the myriad of ways in which that philosophy is put into practice as students investigate phenomena, engage in science and engineering practices, and frame their understanding with the crosscutting concepts. In this chapter, I share how I plan units with science notebooks in mind and give a few examples of how students' work helps me identify their strengths and areas for growth and, as a result, plan instruction accordingly. I hope these examples help you envision how notebooks can fit into your own teaching practice.

Unit Planning With Science Notebooks

Most of my units are organized around a compelling question that is broken down into three or four supporting questions. These questions are generated from a variety of places: phenomena, language from standards (including the *NGSS*, which I consult frequently even though we have not officially adopted them), student-generated questions, and my own goals for the unit. In many cases, I can plan the framework of a unit by anticipating the questions students will pose during our initial investigations. Other times, predetermined questions guide the flow of the unit. The supporting questions help me begin to plan an instructional sequence and identify the types of notebook entries I want to use in the unit. As I think these through, I identify science and engineering practices to emphasize, opportunities for work with crosscutting concepts, and assessment opportunities. I also consider cross-curricular connections, which scientists I can highlight through picture book biographies, and possibilities for guest speakers and field trips. Table 9.1 (p. 106) shows an abbreviated version of the table that guides my planning for the unit.

Chapter 9

Table 9.1. Structure for a fourth-grade unit on Ohio's geologic history using teacher-created questions.

Inquiry unit: Ohio's geologic history			
Compelling question: What can we learn from analyzing Ohio's rocks and fossils?			
Supporting question 1 What can the rock record teach us about the past?	Supporting question 2 How do fossils form? How common are fossils?	Supporting question 3 What kinds of fossils can we find in Ohio?	Supporting question 4 What can we infer about Ohio's past from fossils?
Instructional activities and notebook entries: • Observational drawings of rocks • Sketchnote from informational text and video on rock classification • Explanation of simulated rock strata (layer cake simulation)*	Instructional activities and notebook entries: • Model showing fossil formation • Data table from fossilization game simulation • Argument (CER) answering question: How often do fossils form?*	Instructional activities and notebook entries: • Observational drawings of locally collected fossils (450 mya) • Tally chart and bar graph of identified fossils using field guide (basis for explanation in next column) • Notes about identified organisms	Instructional activities and notebook entries: • Argument answering question: What was the environment in Ohio like 450 million years ago?* • Guided notes (timeline) from reading *Under Ohio: The Story of Ohio's Rocks and Fossils*
Targeted elements of the three dimensions (*NGSS*) • DCIs: ESS1.C-E1** • SEPs: 4-E2; 6-E2; 8-E1 • CCCs: 1-E3; 3-E1	Targeted elements of the three dimensions (*NGSS*) • DCIs: ESS1.C-E1 • SEPs: 2-E4; 4-E2; 7-E4 • CCCs: 7-E2	Targeted elements of the three dimensions (*NGSS*) • DCIs: LS4.A-E1; E2 (grade 3) • SEPs: 4-E1, 8-E4 • CCCs: 1-E3	Targeted elements of the three dimensions (*NGSS*) • DCIs: ESS1.C-E1 • SEPs: 6-E2; 8-E1 • CCCs: 7-E2
Featured scientist N/A	Featured scientist • Mary Anning (*Mary Anning and the Sea Dragon* by Jeannine Atkins; *Dinosaur Lady: The Daring Discoveries of Mary Anning, the First Paleontologist* by Linda Skeers)	Featured scientist • Guest speaker	Featured scientist N/A
Unit assessment: Paragraph answering compelling question and including evidence from investigations, as well as concepts and vocabulary from unit			

Note: This is not a complete list of instructional activities.

*This notebook entry serves as an assessment product for a particular supporting question.

**E indicates the 3–5 grade band, the subsequent number identifies the specific element.

While I try to be as comprehensive as possible in my planning, changes inevitably occur. Either student questions or interests take the class in a different direction than anticipated, or I find that I need to add additional lessons to help students truly understand concepts or vocabulary. I consider this plan to be my starting point and a working document that I update and change over time. Even so, planning in this way allows me to make effective use of science notebooks throughout the unit. How does student work affect my decision-making and revisions to my initial plan? The following case studies provide examples from past years' classes.

Observational Drawing

I introduced this collection of three third-grade student drawings and descriptions of mealworms (Figure 9.1) in Chapter 2. The assignment tasked students with recording observations of three life stages of a mealworm (larvae, pupae, adult) in both sketches and writing. The assignment was purposely open-ended to provide students with opportunities for choice in how they documented their observations. The resulting work gave me great insight into students' proficiencies with these skills and allowed me to personalize feedback and differentiate future instruction.

One student's observation of the pupal form (Figure 9.1a) shows definite strengths along with areas for growth. She clearly understood the benefit of drawing a large picture and had paid close attention to details, including segmentation, the curved and pointed body ("one end is thick and the other is thin"), and tiny protrusions extending from both sides of the body. Additionally, she

Figure 9.1. Third-grade students' observational drawings and descriptions of the life stages of mealworms.

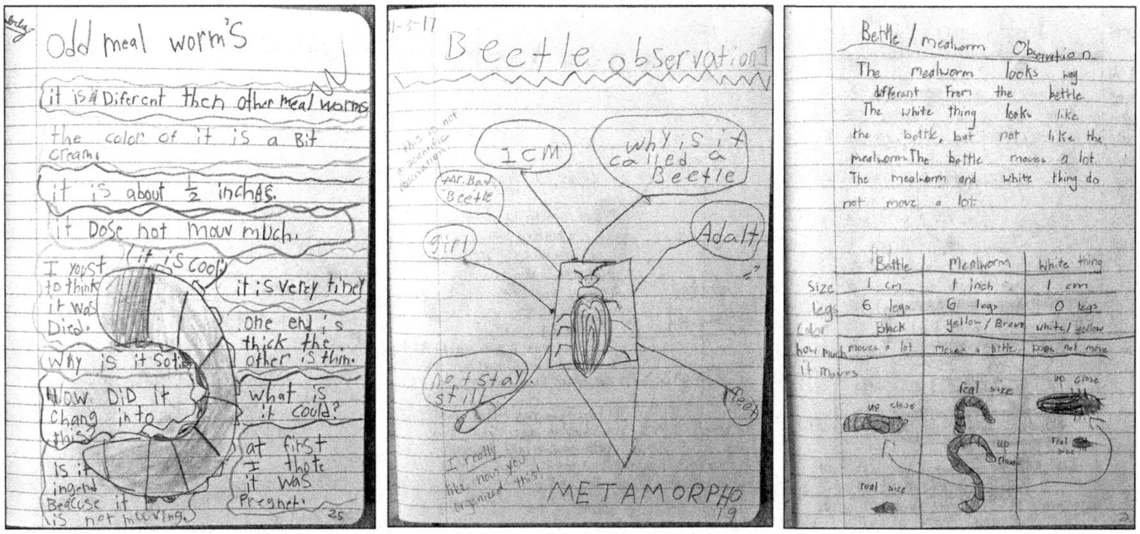

(a) (b) (c)

Observational drawings of the pupal form of a mealworm (a); the adult form of a mealworm (a darkling beetle) (b); and three life cycle stages of a mealworm (note the arrows indicating that two of the student's sketches need to be transposed) (c).

was curious and reflective, using her notebook page to record a number of questions about the mysterious-to-her form ("How did it change into this? What is it called? Why is it so tiny? Is it injured because it is not moving?") as well as her previous thinking ("At first I thought it was pregnant and I used to think it was dead"). As I carefully examined her work, I noticed a lack of adjectives and descriptive phrases in her writing, an instance in which she used subjective language ("It is cool"), and an estimated measurement in customary, not metric, units. I shared these observations with her in conversation and written feedback and made a note to brainstorm a list of adjectives with this student during the next activity involving observation. I also planned a targeted mini-lesson with her and a few other students to review measuring length using the metric system.

Contrast this with a second student's work (Figure 9.1b). Again, this notebook entry shows definite strengths. This student was experimenting with different ways of organizing information, including the web shown in this example. She focused on small details, such as the differing length of the two antennae; the clear head, thorax, and abdomen; and the segmented legs. She also correctly measured the beetle using the metric system of measurement. However, other elements indicate areas for growth, including the smiley face drawn on the beetle's head, the invented name for the beetle ("Mr. Bad Beetle"), descriptions with no supporting evidence (declaring that the beetle is a girl), and an overall lack of written content. This student and I had a one-on-one conversation reviewing the differences between creative illustrations in cartoons and graphic novels and observational drawings used in science and the importance of sticking to factual descriptions. For the next assignment involving observations, I provided her with a set of sentence starters as a scaffold for her writing.

Finally, consider a third student's work (Figure 9.1c). Although she worked from the same directions (draw and describe the different stages of the mealworm's life cycle), she took a remarkably different approach by including all three on the same page in a comparative paragraph and table. Strengths in her work include the comparisons themselves, the appropriate use of a data table to organize related information, and the "real-size" and "up-close" sketches. But as with the others, there are areas for growth as well. Note the inconsistent measurement systems (both customary and metric) and that her "up-close" drawings don't reveal much more detail than her "real-size" ones do. Missing some of these small details captured by her classmates meant that she may not have been able to draw on these sketches to pose the same number and types of questions about these organisms, which was one of our key purposes in recording observations like these. My conversation with this student, as with the other two, shared both these strengths and opportunities for growth. I also made a mental note to find ways to challenge this student and extend her learning through varied questions and tasks in the future.

Students' Experimental Designs

Learning to plan a "fair test" by identifying controls and variables is an important development in the elementary grades (see Chapter 6). Analyzing what my students write and draw in their notebooks during open-ended, collaborative investigations helps me structure lessons to support students in this development. Consider the student work shown in Figures 9.2 and 9.3. During

Figure 9.2. A third-grade student's documentation of an open-ended, collaborative investigation into how cars travel down hills.

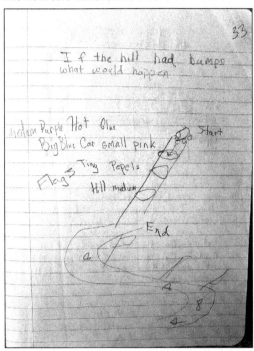

Figure 9.3. The same third-grade student's data table (with the median data point circled) and claim and evidence statement following the experiment, revealing that her data analysis skills are much more advanced than what her documentation of the procedure might suggest.

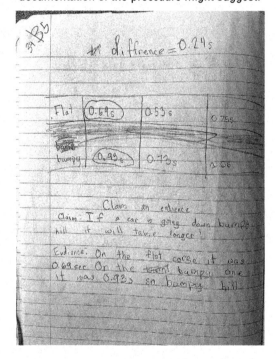

a unit on forces and motion, third-grade students planned and conducted investigations on questions of their choosing related to cars traveling down a hill. Investigation topics varied, with students investigating many different variables, including the incline of the hill and how changing the texture of the hill's surface might affect the car's motion.

One student's group wanted to answer the question "If the hill had bumps, what would happen?" As you can see in Figure 9.2, this student didn't include much in the way of an experimental procedure in her notebook entry, which reveals a need for explicit instruction and practice of experimental design, as well as the importance of accurate documentation. There is a list of materials (including different-size cars that ultimately were not used) and a rudimentary sketch showing a hill with three bumps and a track with flags (to presumably measure distance traveled). No procedures are recorded, which leaves the viewer with many questions: Will the group compare the hill with bumps to a hill without bumps? What data will they collect, and how will they collect these data? Why are different-size cars listed as materials? These are all questions I discussed with this small group as I circulated around the room and monitored student progress. Our conversation revealed that they had a more detailed plan than what was written down and thus were able to move on to actually performing the experiment.

It is quite interesting to contrast this student's documentation of her experimental procedure in Figure 9.2 with her data and claim and evidence statements in Figure 9.3. While the former leaves many questions unanswered, the latter paints a much more complete picture of what happened in that experiment. We can tell that the group tested two hills, one flat and one bumpy; that they measured how long it took the car to travel down the hill; and that they conducted three trials on each hill and used the median data point (circled in the data table). It is not clear from this entry that the hills were equally steep, but this was confirmed in my conversation with the group before the data collection. In my conversation with this group after the investigation was complete, I asked the students to reflect on their entries while I pointed out the questions and observations noted here. We discussed the importance of clearly documenting the materials, controls and variable, and experimental procedure so that others could replicate the investigation, and I planned future lessons on these topics to help them put this feedback into practice. In this particular situation, these themes were evident in the entire class, simplifying planning somewhat. If groups had shown different levels of proficiency in documenting experimental procedures, I would have added more differentiation to my follow-up lessons.

Science notebooks, when used in this student-centered manner, provide a wealth of information that can be used to inform instruction. Thoughtfully reviewing student work in this way is time-consuming, but the opportunities for targeted feedback and student growth are worth the effort. I hope the reflections and examples shared in this book have inspired you to start using science notebooks in your classroom or to maximize their potential if you already do.

Reference

Clarke, C. 2020. Five life lessons from Katherine Johnson, a black mathematician who became a star. *Black Enterprise. www.blackenterprise.com/5-life-lessons-from-the-mathematician-katherine-johnson/.*

APPENDIX

Blackline Masters

Use the blackline masters listed below in your own students' notebooks.

Note: Copy pages 112 and 113 back-to-back (double-sided), ensuring that the pages are oriented correctly. Fold the paper in half to make a booklet. To add to a notebook, glue the blank back of the booklet to the notebook page.

Note: Print or copy this page, then place a sticky note on top of each square. Use the manual feed on your copier to run the page through a second time. This will copy the assessment grid onto each sticky note. The paper can be reused multiple times as a template.

Contents

Page	Investigation

National Science Teaching Association

Appendix: Blackline Masters

Safety Contract

1. I will listen carefully.

2. I will follow directions.

3. I will wash my hands after science activities.

4. I will keep myself and others safe.

5. I will treat all materials with care.

6. I will be a responsible scientist.

_____   _____

Student's Signature   Date

_____   _____

Parent's/Guardian's Signature   Date

_____   _____

Teacher's Signature   Date

Safety Contract

- Listen to the teacher's directions.
- Do not touch, smell, eat, or drink anything unless you are told to do so.
- Be respectful of science materials.
- Wear safety goggles.
- Do not run in the science lab.
- Wash your hands after doing activities.

I have reviewed these safety rules with my teacher and my parent/guardian. I agree to follow these rules and any additional directions given by the school or teacher.

_____   _____

Student's Signature   Date

_____   _____

Parent's/Guardian's Signature   Date

_____   _____

Teacher's Signature   Date

A	B	C
D	E	F
G	H	I
J	K	L

M	N	O

P	Q	R

S	T	U

V	W	XYZ

Word	Definition	Picture
	_____ _____ _____ _____ _____	
	_____ _____ _____ _____ _____ _____	
	_____ _____ _____ _____ _____ _____	
	_____ _____ _____ _____ _____	
	_____ _____ _____ _____ _____	

I See What do you notice?	I Think What do you think about what you see?	I Wonder What do you wonder about what you see?

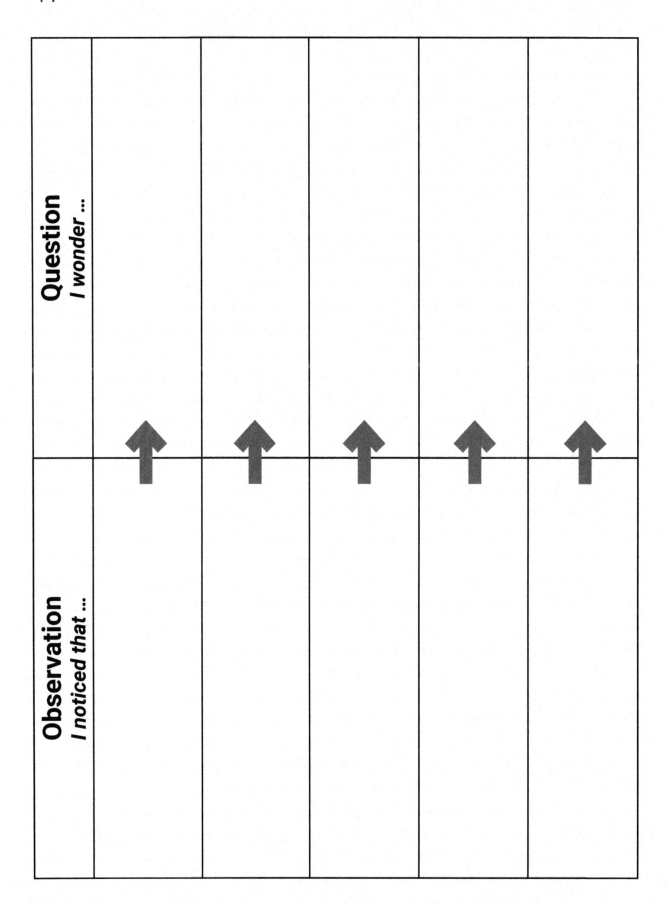

Name: _____

Is my observation **BIG**?	YES	NO
Does it **show details**?	YES	NO
Does it use the **correct colors**?	YES	NO
Is it **fully colored**? (no white space)	YES	NO
Does it have labels?	YES	NO
Is it **neat and organized**?	YES	NO

I can improve by …

Name: _____

Is my observation **BIG**?	YES	NO
Does it **show details**?	YES	NO
Does it use the **correct colors**?	YES	NO
Is it **fully colored**? (no white space)	YES	NO
Does it have labels?	YES	NO
Is it **neat and organized**?	YES	NO

I can improve by …

Name: _____

Is my observation **BIG**?	YES	NO
Does it **show details**?	YES	NO
Does it use the **correct colors**?	YES	NO
Is it **fully colored**? (no white space)	YES	NO
Does it have labels?	YES	NO
Is it **neat and organized**?	YES	NO

I can improve by …

Name: _____

Is my observation **BIG**?	YES	NO
Does it **show details**?	YES	NO
Does it use the **correct colors**?	YES	NO
Is it **fully colored**? (no white space)	YES	NO
Does it have labels?	YES	NO
Is it **neat and organized**?	YES	NO

I can improve by …

Name: _____

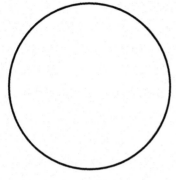

Specimen: _____

Magnification: _____

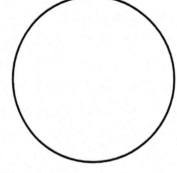

Specimen: _____

Magnification: _____

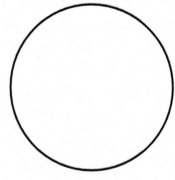

Specimen: _____

Magnification: _____

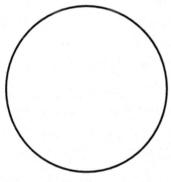

Specimen: _____

Magnification: _____

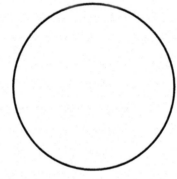

Specimen: _____

Magnification: _____

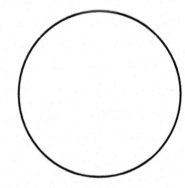

Specimen: _____

Magnification: _____

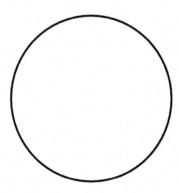

Specimen: _____

Magnification: _____

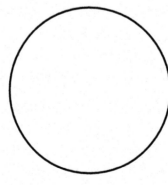

Specimen: _____

Magnification: _____

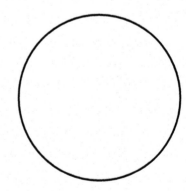

Specimen: _____

Magnification: _____

Variable *What one thing will be changed in your experiment?*	
Controls *What will you keep the same during your experiment?*	
Data *What will you observe or measure to answer your question?*	

Variable *What one thing will be changed in your experiment?*	
Controls *What will you keep the same during your experiment?*	
Data *What will you observe or measure to answer your question?*	

Bar Graph Checklist

☐ The graph has a title.

☐ The intersection is labeled with 0.

☐ The *x*-axis is labeled.

☐ The *y*-axis is labeled.

☐ Numbers on the *y*-axis are labeled on the lines, not in the spaces.

☐ There are equal intervals on the axes.

☐ There is a space between the *y*-axis and the first set of data.

☐ There is equal spacing between sets of data.

☐ If there is a zero amount, it is represented with a line.

☐ Double and triple bars are labeled with a key.

Bar Graph Checklist

☐ The graph has a title.

☐ The intersection is labeled with 0.

☐ The *x*-axis is labeled.

☐ The *y*-axis is labeled.

☐ Numbers on the *y*-axis are labeled on the lines, not in the spaces.

☐ There are equal intervals on the axes.

☐ There is a space between the *y*-axis and the first set of data.

☐ There is equal spacing between sets of data.

☐ If there is a zero amount, it is represented with a line.

☐ Double and triple bars are labeled with a key.

Bar Graph Checklist

☐ The graph has a title.

☐ The intersection is labeled with 0.

☐ The *x*-axis is labeled.

☐ The *y*-axis is labeled.

☐ Numbers on the *y*-axis are labeled on the lines, not in the spaces.

☐ There are equal intervals on the axes.

☐ There is a space between the *y*-axis and the first set of data.

☐ There is equal spacing between sets of data.

☐ If there is a zero amount, it is represented with a line.

☐ Double and triple bars are labeled with a key.

Bar Graph Checklist

☐ The graph has a title.

☐ The intersection is labeled with 0.

☐ The *x*-axis is labeled.

☐ The *y*-axis is labeled.

☐ Numbers on the *y*-axis are labeled on the lines, not in the spaces.

☐ There are equal intervals on the axes.

☐ There is a space between the *y*-axis and the first set of data.

☐ There is equal spacing between sets of data.

☐ If there is a zero amount, it is represented with a line.

☐ Double and triple bars are labeled with a key.

Writing and Talking About Science

ASKING A QUESTION

How are _____ and _____ alike?

How are _____ and _____ different?

What would happen if _____?

I wonder _____.

OBSERVING

I noticed _____.

I saw/heard/felt/smelled _____.

PREDICTING

I think _____ because _____.

I predict _____ because _____.

If _____ , then _____ because _____.

ANALYZING DATA

The data show that _____.

The graph shows that _____.

A pattern in the data is _____.

Writing and Talking About Science

MAKING A CLAIM

I claim that _____.

My evidence is _____.

The evidence from _____ shows that _____.

The reason this evidence supports the claim is _____.

This evidence does not support the claim because _____.

READING/WATCHING A VIDEO

A question I have is _____.

I learned that _____.

This connects to our investigation because _____.

This supports my idea because _____.

This does not support my idea because _____.

Task List

☐ _____

☐ _____

☐ _____

· ·

Task List

☐ _____

☐ _____

☐ _____

· ·

Task List

☐ _____

☐ _____

☐ _____

	+	✓	−
Clear			
Complete			
Accurate			
Uses science words			
Uses fact words			

	+	✓	−
Clear			
Complete			
Accurate			
Uses science words			
Uses fact words			

	+	✓	−
Clear			
Complete			
Accurate			
Uses science words			
Uses fact words			

	+	✓	−
Clear			
Complete			
Accurate			
Uses science words			
Uses fact words			

	+	✓	−
Clear			
Complete			
Accurate			
Uses science words			
Uses fact words			

	+	✓	−
Clear			
Complete			
Accurate			
Uses science words			
Uses fact words			

Name: _____

		YES	NO
Used the bar graph checklist while working.			
☐	The graph has a title.		
☐	The intersection is labeled with 0.		
☐	The x-axis is labeled.		
☐	The y-axis is labeled.		
☐	Numbers on the y-axis are labeled on the lines, not in the spaces.		
☐	There are equal intervals on the axes.		
☐	There is a space between the y-axis and the first set of data.		
☐	There is equal spacing between sets of data.		
☐	If there is a zero amount, it is represented with a line.		
☐	Double and triple bars are labeled with a key.		

Student Name: _____

Share your Science Notebook with your grown-ups!

Return your notebook on _____.

I'm proud of bookmarked page _____ because

_____

_____

I'm proud of bookmarked page _____ because

_____

_____

On notebook page _____ , I learned something new. Here's what I learned!

_____

_____

Something I want to get better at is _____

_____

_____

My child shared their Science Notebook with me, and I noticed _____

_____

_____

_____

Grown-up's Signature _____

Student Name: _____

Share your Science Notebook with your grown-ups!

Return your notebook on _____.

I'm proud of bookmarked page _____ because

_____

_____

I'm proud of bookmarked page _____ because

_____

_____

On notebook page _____ , I learned something new. Here's what I learned!

_____

_____

Something I want to get better at is _____

_____

_____

My child shared their Science Notebook with me, and I noticed _____

_____

_____

_____

Grown-up's Signature _____

Index

Page numbers printed in **boldface type** indicate figures or tables.

Index

Index

Index

Index

National Science Teaching Association